Relevant Applications of High-Performance Liquid Chromatography in Food, Environmental, Clinical and Biological Fields

Edited by Oscar Núñez

Published in London, United Kingdom

Relevant Applications of High-Performance Liquid Chromatography in Food, Environmental, Clinical and Biological Fields
http://dx.doi.org/10.5772/intechopen.1003433
Edited by Oscar Núñez

Contributors
Ali Gamal Al-Kaf, Ali Hakimi, Amin Alaei, Ariaya Hymete, Ayenew Ashenef, Burhan Ceylan, Clara Oliva Mussarra, Esayas Tesfaye, Fatemeh Keyfi, Feruza Ahmed, Goran Mitulović, Hadiseh Mirahmadi Daryasari, Hend Ahmed Al-Haj, Jadranka Miletić Vukajlović, Judit García-Villoria, Pouya Gharavi, Rana Abdullah Al-Robaidi, Robert Owen Bussey III, Sixue Chen, Tadios Niguss, Tekeste Abebe, Yatendra Singh

First published in London, United Kingdom, 2025 by IntechOpen
IntechOpen is the global imprint of INTECHOPEN LIMITED, registered in England and Wales, registration number: 11086078, 167-169 Great Portland Street, London, W1W 5PF, United Kingdom

For EU product safety concerns: IN TECH d.o.o., Prolaz Marije Krucifikse Kozulić 3, 51000 Rijeka, Croatia, info@intechopen.com or visit our website at intechopen.com.

British Library Cataloguing-in-Publication Data
A catalogue record for this book is available from the British Library

Relevant Applications of High-Performance Liquid Chromatography in Food, Environmental, Clinical and Biological Fields
Edited by Oscar Núñez
p. cm.
Print ISBN 978-0-85466-122-0
Online ISBN 978-0-85466-121-3
eBook (PDF) ISBN 978-0-85466-123-7

If disposing of this product, please recycle the paper responsibly.

Meet the editor

Oscar Núñez studied chemistry at the University of Barcelona in Spain, where he also received his Ph.D. in 2004. He has been a Full Professor in the Section of Analytical Chemistry at the University of Barcelona since 2020. He has more than 165 scientific papers and book chapters to his credit. He has also edited eight books on liquid chromatography, liquid chromatography-tandem mass spectrometry (LC-MS/MS), sample preparation techniques in food analysis, capillary electrophoresis, and food integrity and authenticity. He has extensive experience in developing liquid chromatography methods with ultraviolet and fluorescence detection, liquid chromatography coupled with low- and high-resolution mass spectrometry, and in sample treatment procedures for environmental and food analysis. His primary research areas include the characterization, classification, and authentication of food and natural products and the prevention of food fraud.

Contents

Preface

The demand for high-throughput separations in food, environmental, clinical, and bioanalytical research is rising, driven by the need to accurately identify and qualify a wide range of compounds in complex matrices. In this sense, high-performance liquid chromatography (HPLC) has become one of the most widely used and well-established separation techniques to meet these demands. This is thanks to its great versatility in chromatographic separation modes (e.g., reversed-phase, normal-phase, HILIC, ion-chromatography, multidimensional-chromatography), diverse column technologies (conventional HPLC columns, sub-2 μm UHPLC columns, partially porous core-shell columns, monolithic columns), and various detection systems (ultraviolet-visible, fluorescence, amperometric). Coupled with both low-resolution and high-resolution mass spectrometry, HPLC stands out as an effective solution for addressing emerging analytical challenges across a broad spectrum of applications.

This book provides an overview of HPLC's advances and new applications in different fields. It is organized into two sections.

Section 1, "Principles and New Advances in General HPLC Applications", contains four chapters. In Chapter 1, "Applications of Ultra High-Performance Liquid Chromatography Coupled with High-Resolution Mass Spectrometry", Bussey III addresses the role of UHPLC coupled to HRMS in food and environmental applications, highlighting the benefits and limitations of HRMS. The chapter also presents three detailed case studies. In Chapter 2, "Optimizing Ultra-High-Performance Liquid Chromatography (UHPLC) Methods for Rapid Analysis of Complex Biological Samples", Hymete et al. explore different strategies for optimizing UHPLC methods aiming to achieve rapid and reliable analyses. The authors consider key parameters such as column selection (including sub-2 μm particle column technology), mobile phase composition, flow rate, and sample preparation, providing valuable information for researchers seeking to leverage UHPLC to efficiently analyse complex biological samples.

The trend toward instrument miniaturization in recent years has made it possible to develop new and sophisticated analytical techniques, such as nano-liquid chromatography (nano-LC). This has made it possible to improve the sensitivity and resolution of chromatography. Nano-HPLC has become essential for both qualitative and quantitative methods. Thus, in Chapter 3, "Theory and Recent Applications of Nano-Liquid Chromatography", Tesfaye et al. cover the theoretical aspects of nano-LC and its current practical uses in analysing pharmaceutical and biological molecules. Furthermore, the authors examine future prospects for the development of nano-LC techniques. Similarly, multidimensional separation systems offer several advantages over traditional one-dimensional separation systems, particularly their ability to separate molecules from complex mixtures. Two-dimensional liquid chromatography (2D-LC) significantly enhances the ability to analyze complex mixtures by providing greater separation power, sensitivity, and flexibility, making it an invaluable tool for metabolomics research. The 2D-LC is an exciting mode when pursuing untargeted analysis, as it allows for high-resolution separation and subsequent identification and quantification of more analytes. In Chapter 4, "Two-Dimensional Liquid Chromatography Advancing Metabolomics Research", Singh and Chen summarize the current applications of 2D-LC in metabolomics and the setups of different separation modes that are being employed, presenting the most suitable combinations of chromatographic methods for different targeted and untargeted metabolomics applications.

Section 2, "Specific HPLC Applications", includes five chapters, three of which are devoted to clinical HPLC applications. In Chapter 5, "Utilizing High-Performance Liquid Chromatography (HPLC) in Clinical Diagnostics", Keyfi et al. advocate for strategically incorporating HPLC technology into clinical diagnostic protocols, ensuring improved patient outcomes through enhanced diagnostic precision and efficiency. They propose integrating HPLC into clinical diagnostic procedures to enhance accuracy, sensitivity, and disease detection and monitoring efficiency. By leveraging HPLC's capabilities, they aim to streamline the analysis of biomarkers, drugs, metabolites, and toxins in various bodily fluids, facilitating rapid and reliable diagnosis of diseases ranging from metabolic disorders to cancers. Moreover, HPLC's ability to analyze multiple analytes simultaneously can expedite patient assessment and treatment decision-making processes. In Chapter 6, "High-Throughput Chromatography for Clinical Proteomics Applications", Mitulović and Vukajlović address the use of high-throughput analysis of clinical samples to monitor the progression or regression of disease or the concentration of therapeutic agents during personalized treatment. In this sense, the use of micro and capillary chromatography is well established, but nano-LC can provide significantly higher sensitivity, especially for analyzing therapeutic monoclonal antibodies. Using new separation columns (e.g., monolithic columns) in nano-LC can provide both high sample throughput and high sensitivity. Additionally, when hyphenated with a high-end mass spectrometer, the combination can be used for screening and targeted analyses of clinical samples.

HPLC stands as a robust analytical technique with extensive applications in clinical diagnostics. Its versatility enables the precise separation, identification, and quantification of complex mixtures in biological samples. In Chapter 7, "Metabolomic Diagnostic in Inherited Metabolic Disorders: Historical Progress and Future Perspectives", Musarra and García-Villoria focus on inherited metabolic disorders (IMDs), which encompass a diverse range of monogenic disorders disrupting metabolic pathways, leading to significant morbidity and mortality. While some treatable IMDs are part of newborn screening programs, the majority remain challenging to diagnose early. There is a pressing need for untargeted metabolomics to address these challenges—a holistic approach using HPLC and high-resolution MS to measure thousands of metabolites in a single analysis. This approach promises to revolutionize diagnostics by enabling comprehensive metabolite detection, optimizing resources, and streamlining diagnostic workflows.

The last two chapters of this section are devoted to applications of HPLC techniques in herbal products. In Chapter 8, "Advances in Applications of High-Performance Liquid Chromatography in the Analysis of Herbal Products", Al-Kaf et al. focus on the utilization of HPLC techniques for the analysis of herbal products by describing step-by-step protocols for the chemical profiling or fingerprinting of herbs and herbal mixtures. Finally, Chapter 9, "A New HPLC-DAD Method for the Simultaneous Measurements of Glycyrrhizic Acid (GA) and Glabridin (GB) in Licorice (*Glycyrrhiza glabra* L.) Extract", by Ceylan, is devoted to developing an easy-to-use, fast, and economical HPLC-DAD method to analyze GA and GB in products containing extracts from *Glycyrrhiza glabra* L. (Licorice), a plant with important applications in the fields of medicine, food, and cosmetics.

We hope this book will prove to be a valuable and useful resource for those working with HPLC analytical methods and their application in different fields.

Oscar Núñez
Department of Chemical Engineering and Analytical Chemistry,
University of Barcelona,
Barcelona, Spain

Section 1

Principles and New Advances in General HPLC Applications

Chapter 1

Applications of Ultra High-Performance Liquid Chromatography Coupled with High-Resolution Mass Spectrometry

Robert Owen Bussey III

Abstract

Ultra High-Performance Liquid Chromatography (UHPLC) coupled with High-Resolution Mass Spectrometry (HRMS) systems allow researchers to potentially identify and quantitate semi-volatile and nonvolatile chemical constituents in aqueous samples. High-resolution mass spectrometers usually show parent compounds and their fragments with mass to charge ratios with uncertainties in the fifth decimal place, whereas low-resolution mass spectrometers only show nominal masses with uncertainties to the first decimal place. This extra information allows for more descriptive data to characterize compounds such as the chemical formula to both the parent compound and the fragments. This chapter will describe the basic principles behind coupling UHPLC and HRMS along with the benefits and potential limitations. The different types of high-resolution mass spectrometers will be described. In addition, this chapter will review several case studies demonstrating how UHPLC coupled with HRMS systems are used in the environmental and food analysis fields along with the application of metabolomics to some of these fields.

Keywords: ultra high-performance liquid chromatography, UHPLC, high-resolution mass spectrometry, HRMS, environmental analysis, food analysis, metabolomics

1. Introduction

This chapter focuses on the application of Ultra High-Performance Liquid Chromatography (UHPLC) coupled with High-Resolution Mass Spectrometry (HRMS). UHPLC is a technique that allows for the separation and analysis of complex mixtures by polarity and column elution time. Compounds that are amenable to UHPLC separation are normally semi-volatile and nonvolatile compounds [1]. As the compounds elute from a column with some degree of chromatographic separation, the compounds are ionized in an ionization source, and then can be coupled to a mass spectrometer to measure the mass spectra of each separated compound. The mass spectra consist of a mass to charge (*m/z*) ratio of a parent compounds or possibly fragments. Mass spectrometers can have a low or high mass resolution based on the ability to differentiate ions that have small mass to charge differences. The accuracy of the *m/z* values is much higher in high-resolution mass spectrometers with *m/z* uncertainties measured to the fifth decimal place versus the first decimal place in low-resolution mass spectrometers [1]. The chromatographic separation and the high mass accuracy complement each other to balance sensitivity versus selectivity to give the most accurate structural information with possible amounts present.

Two of the most used HRMS instruments are Time of Flight (TOF) and orbitrap systems. To increase the selectivity of these HRMS systems, quadrupoles are often integrated to create hybrid systems that can filter for *m/z* ranges or specific *m/z* ratios, and these systems contain the "Q" designation for quadrupole [1]. HRMS systems will be used in this chapter to refer to the Q-TOF or Q-Orbitrap hybrid, high-resolution mass spectrometers. In addition, the benefits and limitations of HRMS systems will be discussed in the following areas: elemental composition elucidation, mass calibration requirements, and chromatographic peak resolution.

The case studies used in this chapter show how UHPLC coupled with HRMS systems are used to analytically deduce complex environmental and food samples along with how metabolomics is used to help simplify the process. Case Study 1 uses HRMS to monitor specific antibiotics and breakdown products in fish grown in artificial aquaculture conditions. Case Study 2 creates a metabolomics workflow with an all-ion fragmentation strategy for an in-house database for common biologically relevant compounds in urine, plasma, and cell lysate. This workflow could be applied to unknown compounds in the future if implemented in a non-targeted analysis. Case Study 3 describes a new method for detecting, measuring, and evaluating known and unknown per- and polyfluoroakyl substances in industrial wastewater analysis. All case studies should show how UHPLC and HRMS are integrated together to add value to scientific studies.

2. High-resolution mass spectrometry

2.1 What is low- versus high-resolution?

Mass spectrometers measure the *m/z* of chemical compounds which means that the compound must be an ion or a charged species to be visualized by a mass spectrometer [1]. An ionization source applies that charge to compounds before they enter the mass spectrometer. At times, it is hard to differentiate between peaks with the same mass, different masses, or different isotopic species at the same retention time. Chromatography increases the accuracy of measuring the correct mass or the accurate mass [2]. More will be discussed later how chromatographic separation can affect accurate mass determination, but when chromatography cannot solve this issue, understanding the mass resolutions of the mass spectrometer becomes important.

Mass resolution is the ability of the mass spectrometer/analyzer to differentiate signals of two ions with small differences in their *m/z* values [2, 3]. Calculating the mass resolution is beyond the scope of this chapter and will not be discussed. Low-resolution mass spectrometers measure nominal mass and usually measure the *m/z* values to the whole number with uncertainty in the first decimal place [1, 2]. The nominal mass is the mass calculated using the most abundant elemental isotopes for

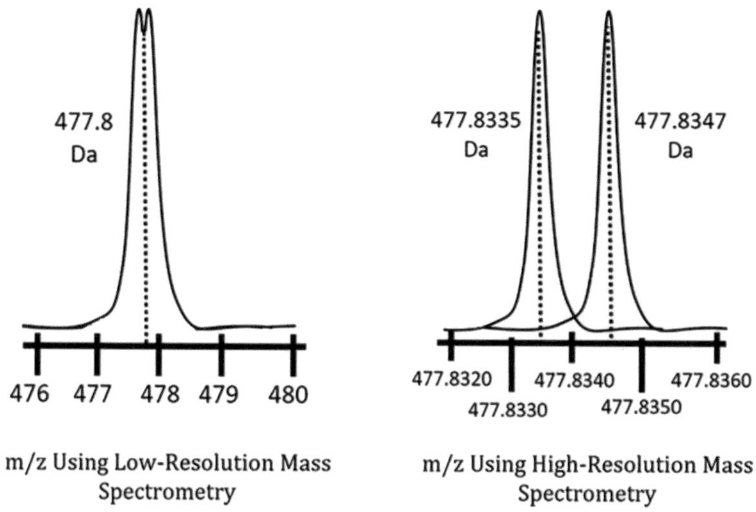

Figure 1.
Difference in the resolving power between low- and high-resolution mass spectrometry. High-resolution can resolve the peaks and give accurate mass.

each atom in the organic formula i.e., ^{12}C isotope versus ^{13}C. Whereas high-resolution allows for measuring the accurate mass *m/z* which can be correlated to a calculated exact mass or the *m/z* of specific isotopic distributions, not just the most abundant [1, 2]. The accurate mass can be measured to four to six decimal places with high-resolution mass spectrometry [1]. **Figure 1** demonstrates the nominal mass versus accurate mass of peaks that were not resolved in low-resolution mass spectrometry, and then were resolved with high-resolution mass spectrometry.

2.2 Basic high-resolution mass spectrometer structure

This section will focus on introducing quadrupole time of flight and quadrupole orbitrap high-resolution mass spectrometers. While the highest mass resolution can be obtained through Fourier-Transform Ion Cyclotron Resonance (FTICR), the FTICR sacrifices analyte sensitivity for better mass resolution [1–3]. Details of FTICR will not be discussed in this chapter and are only mentioned as an option if higher mass resolution is needed in an analysis.

Quadrupole Time of Flight mass spectrometers are hybrid HRMS systems that measure the *m/z* using the travel time of the ions in a set path length [4]. As can be seen in **Figure 2**, the ion path starts as the sample exits the UHPLC as chromatographically separated compounds. As each compound arrives at the ionization source, they are ionized, enter the front of the mass spectrometer, pass various components like a quadrupole filter, and then enter the Time-of-Flight chamber. The ions are accelerated through an electric field by the ion pusher and then pass through electrodes called a reflectron that focus the ions, reflect the ions to change direction, and maintain ion velocity in the new direction [4]. As can be seen in **Figure 2**, the ion path lengths can have multiple variations such as a V, W, or N (not pictured) which can change the mass resolution [4]. While all ions are given the same energy along the path length, smaller ions hit the detector first and larger ions hit the detector later since the high mass ions move slower [4]. With instrument calibration, the *m/z* is measured, and

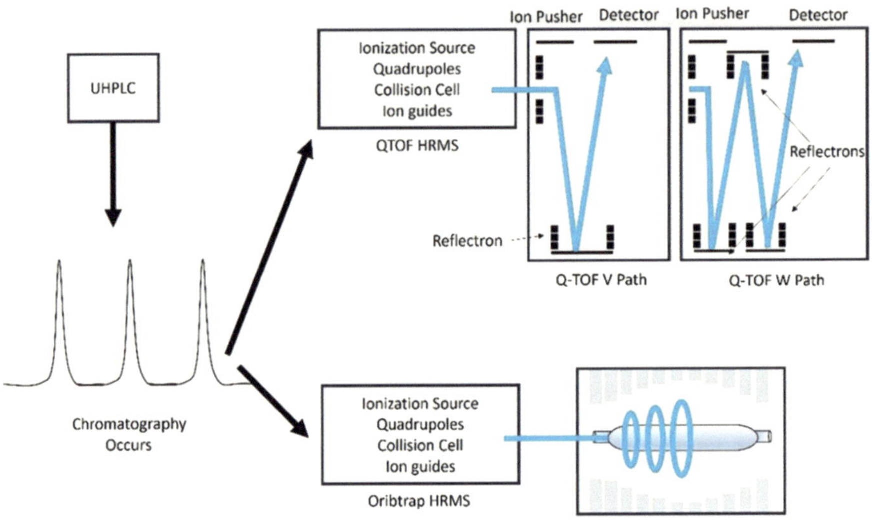

Figure 2.
Sample and ion path into high-resolution Q-TOF and Q-Orbitrap mass spectrometers.

generally higher mass resolution can be achieved with longer ion path lengths (N and W paths), but often at the expense of sensitivity of the signals.

Quadrupole orbitrap mass spectrometer cyclotron analyzers are made up of a central electrode in the shape of a wide rod. The rod is surrounded by an outer electrode as shown in **Figure 2**. The ion path starts as the sample exits the UHPLC as chromatographically separated compounds. As each compound arrives at the ionization source, they are ionized, enter the front of the mass spectrometer, pass various components, and then enter the cyclotron inside the orbitrap. The cyclotron consists of a central electrode where the ions orbit around continuously [5]. The housing that contains the inner electrode is called the outer electrode and helps the ions orbit around the central electrode. As the ions orbit, they can be separated because the orbital frequency is usually inversely related to m/z. The ions are detected through image current electrodes [6].

2.3 Benefits and limitations of high-resolution mass spectrometry

This section will go over a list of benefits and potential limitations of high-resolution mass spectrometry. This list may not be comprehensive, but these are the most important points while planning an experiment that may require HRMS like the Q-TOF or the Q-Orbitrap.

2.3.1 Elemental composition assignment

Some key terms to remember when understanding elemental composition assignment with high-resolution mass spectrometry are as follows: exact mass, accurate mass, mass accuracy, and mass defect. In the previous sections, exact mass and accurate mass were explained, but not in relation to determining the elemental composition assignment. The accurate mass is experimentally determined with HRMS systems and can differentiate compounds with the same nominal mass [1]. The accurate mass is then used in combination with the calculated exact mass (mass with specific isotopic distribution) to help determine the elemental composition of both the parent ion and ion fragments. The difference between the accurate mass and exact mass is the mass accuracy and is usually measured as a difference relative to the exact mass multiplied by 1E6 to give ppm [1]. The lower the mass accuracy value, the better the confidence there is in a suggested elemental composition calculation [1, 2]. In addition, HRMS systems measure masses so accurately that the elemental composition/isotopic distribution can then be further elucidated by using the characteristic mass defect [2]. The mass defect is the difference in the nominal mass (most abundant isotopic distribution) and the exact mass [2].

A limitation in determining elemental composition is the mass range. As the mass of a compound increases, mass resolution also must increase to maintain mass accuracy. The main problem is that with higher masses, there are more possible combinations of the chemical formula that match the criteria. This requires further fragmentation experiments to narrow down the correct formula [2].

2.3.2 Maintaining mass accuracy with HRMS with mass calibration

The high mass accuracy is not just a default setting for high-resolution mass spectrometers. Work must be done to perform external mass calibration to create trusted accurate mass measurements with a mixture of reference compounds and storing those reference values as instrument parameters in the software [1]. To streamline instrument calibration, reference standards are often available pre-mixed from the mass spectrometer manufacturers and the calibration process is automated or semi-automated in the software [1]. External calibration is not required as often contrary to internal calibration required for every experiment. Often during experiments, mass accuracy can drift leading to variations in mass accuracy calculations and structural elucidation [1]. Internal calibration uses a secondary pump and an ionization source baffle to always introduce the mass calibrant into the mass spectrometer source with the baffle as the controller for the mass spectrometer inlet. The software helps to alternatively change the baffle position to allow for sample and internal calibrant spray to the mass spectrometer inlet. The internal mass calibrant helps refine the mass accuracy when there are variations in the mass spectrometer's environmental conditions and can be considered required or optional depending on the manufacturer. The software often automates the internal mass calibration calculations [1].

2.3.3 Chromatographic peak resolution troubleshooting

Often the mass accuracy of the HRMS can be affected by outside influences such as issues with UHPLC separations or mass spectrometer health. When chromatographic peaks are not symmetrical or there is an increase in signal to noise, the mass accuracy of the HRMS is not reliable because the accurate mass calculation changes due to the peak apex [2]. Frequent issues which can impact the mass accuracy include the liquid chromatography separation not resolving peaks, a contaminated liquid chromatography system causing unique peaks shape changes/baselines shifts, or an issue with the mass spectrometer causing the same change in peak shape to all peaks. Re-running the sample after cleaning and recalibrating both the UHPLC and the HRMS system should help assign the cause to the UHPLC or the mass spectrometer. If the chromatographic separation has co-eluting peaks, the two co-eluting peaks will have an accurate mass average not exactly representative of either peak. In this instance, the UHPLC parameters need to be changed or a higher resolution needs to be obtained to resolve the peaks. It is important to note that increasing the resolution does not always improve the mass accuracy because sensitivity drops [2].

3. Case studies

3.1 Case study 1: antibiotic and breakdown product determination in fish raised in aquaculture environments using HRMS

3.1.1 Background

The use of antibiotics to prevent disease in fish raised in aquaculture environments has become more prevalent in the world as growing fish and other livestock in small artificial environments has become a popular technique for protein production [7]. Often when animals are grown in these environments bacterial infections can be prevalent. To prophylactically prevent bacterial infection, antibiotics are often misused in large quantities causing excess runoff into environmental water sources and leading to bioaccumulation in animals [7, 8]. When antibiotics are in high abundance in aquacultural food sources and the environmental water sources, antibiotic resistant strains of bacteria can emerge which in turn can increase mortality in the human population [7, 9].

Enrofloxacin is a fluoroquinolone antibiotic that acts to inhibit DNA replication in bacteria and is often overused in the aquaculture environment [10]. When modified by the bacteria, enrofloxacin becomes another antibiotic called ciprofloxacin (**Figure 3**); other metabolites can also form which all may have antibacterial properties or other toxicity issues [11]. Dai et al. created a targeted/non-targeted high-resolution mass spectrometry method for fish tissue to monitor these specific antibiotics and other breakdown products that currently are unknown [11]. These unknown metabolites could then be screened by regulatory agencies in the future.

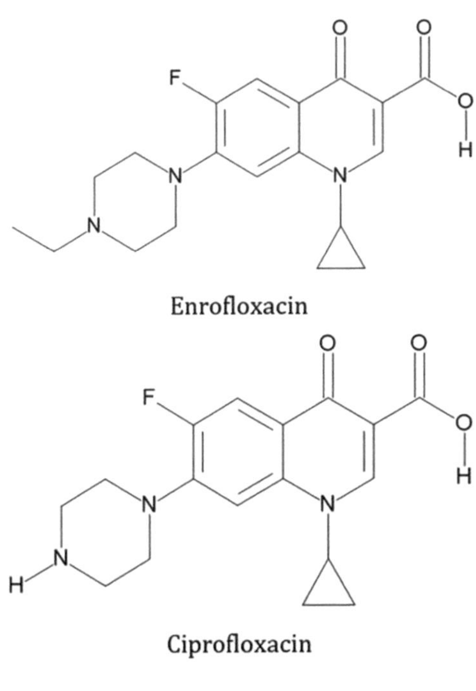

Figure 3.
Chemical structures of enrofloxacin and ciprofloxacin.

3.1.2 Instrumentation and method

The liquid chromatography method used a Thermo Ultimate 3000 UHPLC system with a Hypersil GOLD (150 mm X 2.1 mm, 3 µm) at 40°C. The liquid chromatography solvents consisted of water with 0.15% formic acid (Solvent A) and methanol (Solvent B) with a flow at 0.3 mL per minute. The gradient included an initial 25% solvent B with a 1-minute hold, an increase to 90% solvent B in between 1 and 5 minutes, a hold at 90% B between 5 and 8 minutes and then re-equilibration to initial conditions for a total of 14 minutes. This method was optimized to decrease the co-elution from unknown metabolites in the fish matrix.

The mass spectrometer used was a Thermo Q-Exactive quadrupole Orbitrap high-resolution mass spectrometer with an electrospray positive, ionization source. The following parameters were optimized for qualitative and quantitative analysis of enrofloxacin accurate mass of 360.1718 Da and ciprofloxacin accurate mass of 332.1405 Da (both M + H ions which is parent plus a proton): spray voltage 3.5 kV, capillary temperature 320°C, vaporizer temperature of 350°C, full MS/dd-MS^2 mode with resolution at 70,000 and 17,500 for each mode (full scan from 150 to 700 Da with data-dependent fragmentation of five most abundant ions at each retention time), normalized energy of 20 and 45 for full scan and ddMS^2.

3.1.3 Sample information and sample preparation

The overall targeted/non-targeted investigation of the enrofloxacin metabolites used a control fish that tested negative for enrofloxacin (never exposed to enrofloxacin) and farmed crucian carp that were bought from the market with no antibiotic-free guarantee. The fish samples were homogenized in a blender and were frozen for future analysis. Standards of enrofloxacin and ciprofloxacin were made along with deuterated enrofloxacin and ciprofloxacin as internal standards.

In total, fifteen fish tissues were obtained from the market. Once analysis was ready to be performed, the fish tissue was thawed, underwent maceration, sonication, solvent extraction, centrifugation, rotary evaporation, reconstitution, and then solvent partitioning. Identity of the presence or absence of the enrofloxacin and ciprofloxacin was done based on the retention time, accurate mass of the parent ion ($M + H^+$ or ion before fragmentation), and accurate mass of the MS^2 spectrum fragments. The peak areas of the parent masses of enrofloxacin and ciprofloxacin were used along with calibration curves to quantitate the amount in the fish samples. Validation protocols showed good repeatability, extraction efficiencies, and low matrix interferences.

3.1.4 Results

Compound Discoverer software was used to compare chromatographic features in both the negative control fish (antibiotic-free guarantee) to the fish from the market. With the help of the comparison function, spectral database searches, elemental composition, and degradation predictor, Dai et al. was able to confirm the presence of enrofloxacin, its metabolite ciprofloxacin, and another unknown metabolite in the fish and not the antibiotic-free fish [11]. With the help of the software and the accurate mass of both the parent and fragments, the unknown metabolite was determined to be desethylene enrofloxacin (**Figure 4**). This analyte was also present in previous research of chicken meat and can be seen as both an enrofloxacin metabolite and a

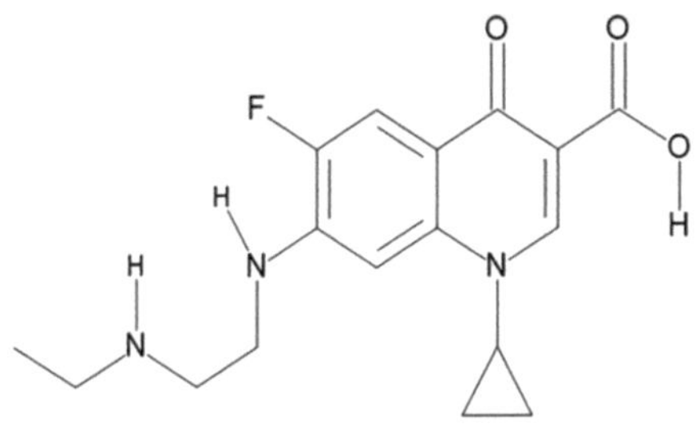

Desethylene enrofloxacin

Figure 4.
Chemical structure of desethylene enrofloxacin.

compound that may change toxicity [12, 13]. Desethylene enrofloxacin underwent semi-quantitation with the deuterated enrofloxacin internal standard since a pure standard could not be obtained.

Enrofloxacin was present in fourteen of fifteen samples tested from the market. Ciprofloxacin and desethyelene enrofloxacin were present in twelve out of the fifteen samples tested. One sample out of the group did not have enrofloxacin, ciprofloxacin, or desethylene enrofloxacin. Three out of the fifteen samples had estimated antibiotic/metabolite sums potentially above the Chinese regulatory limit [11]. Using UHPLC and HRMS, this study showed that it could successfully quantitate enrofloxacin and ciprofloxacin in fish from aquaculture. In addition, the high-resolution scan of both the abundant parent compounds and fragments at each retention time allowed for screening of new metabolites in aquaculture. Knowing which metabolites are present allows for a more informed decision on the content and toxicity of the food samples.

3.2 Case study 2: development of HRMS metabolomics workflow using all ion fragmentation

3.2.1 Background

Metabolomics has been used in many fields to map complex pathways or characterize complex mixtures, which require discreet separation and high mass accuracy to effectively identify compounds. The more descriptive data that can be added to an analyte of interest, the higher the confidence in the compound identification for matching against potential mass spectral databases [14, 15]. High-resolution mass spectrometry can improve the accuracy of metabolite identification, and researchers have explored ways to use HRMS to identify compounds of interest [16]. This work by Naz et al. creates a HRMS metabolomics workflow to help annotate and add descriptive data to analytes that may be biologically relevant markers in urine, plasma, or cell lysate [17]. Having a better way to add descriptive data to identify and differentiate analytes of interest in large metabolomics datasets enhances data interpretation and accuracy in metabolite identification.

Certain metabolomic requirements such as accurate mass, retention time, isotopic patterns, and MS spectrum were suggested in 2007 as part of the Metabolite Standard Initiative [18]. The goal of the workflow presented by Naz was to create screening methods that can add even more descriptive data to increase confidence and level of identification as can be seen in **Figure 5**. This screening

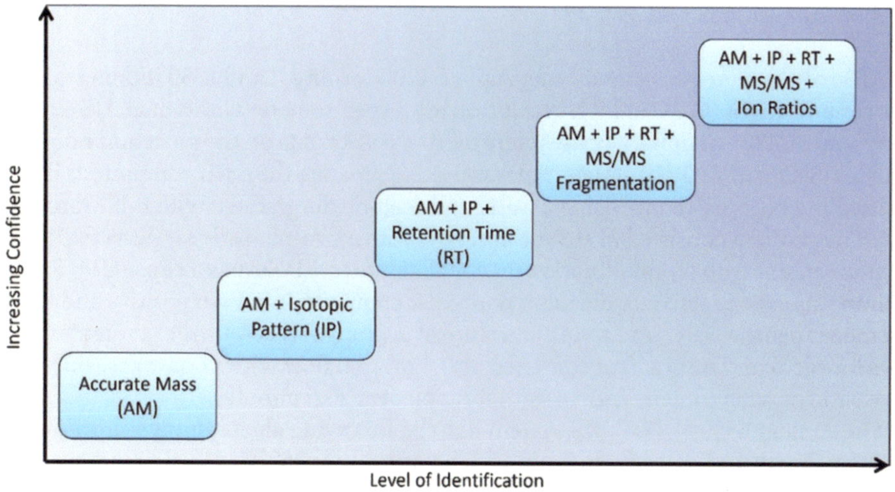

Figure 5.
Level of confidence versus level of identification in metabolomics workflow.

method incorporates higher confidence identifications within targeted in-house databases and increases confidence for new compounds not included in those databases. New compound differentiation is important because in-house databases may or may not have the same extensive list of chemical classes as in commercial databases.

Using HRMS with a good separation strategy to separate and resolve chemical classes is important in metabolomics. This workflow includes both reversed phase (RP) and hydrophilic interaction liquid chromatography (HILIC) to chromatographically resolve as many chemical classes as possible. When coeluting isobaric constituents (sharing the same mass to charge ratio) cannot be separated, HRMS can help differentiate these compounds.

All-ion fragmentation, a derivation of Data-Independent Acquisition (DIA), was suggested as an improvement to non-targeted analysis HRMS workflows involving both parent and fragments. The all-ion fragmentation mode includes the collection of three scan measurements with 0, 10, and 30 eV collision energies at each data point. Collecting multiple fragmentation energies increases the chance of detecting both parent and fragments of most compounds in the database. This information can then be used to investigate the identity of the compounds and quantify the compounds.

Non-targeted, metabolomic studies require quantification to allow for reporting of relative levels of compounds present. In this study, the peak area from the Extracted Ion Chromatogram (EIC) of either the parent or fragment ions is used as the quantifying ion using calibration curves of reference compounds. The other fragment ion EIC peak areas are used as the qualifier ions. In addition to quantification, the peak areas can be put into ratios of the qualifier/quantifier for identification. Data is created for neat ratios and ratios in plasma, urine, and HCT116 cell lysate matrices. While this workflow can be used for in-house databases, it can also be used for compounds that are not found in the database. Data can be compared to known versus unknown compounds to elucidate the possible structure and relative levels.

3.2.2 Instrumentation and method

All analyses were performed on an Agilent 1290 Infinity II and 6550 iFunnel quadrupole time-of-flight (Q-TOF) high-resolution mass spectrometer with a dual AJS electrospray source. Two stationary phases were used to collect data on the polar and non-polar analytes. SeQuant ZIC-HILIC column was used to separate the polar compounds in both positive and negative modes using a water and acetonitrile gradient with 0.1% formic acid. The gradient consisted of 0.3 ml/min flow with a two-phase linear decrease in organic content with re-equilibration over 40.5 minutes. RP Zorbax Eclipse C18, RRHD column was used to separate the more non-polar compounds in both positive and negative modes using a water and a 90:10 mixture of 2-propanol:acetonitrile gradient with 0.1% formic acid. The gradient consisted of 0.4 ml/min flow with a one-phase linear increase in organic content with re-equilibration over 20.5 minutes.

The Agilent 6550 Q-TOF-MS system had the following electrospray ionization settings: sheath and drying gas 8 and 15 L per minute at 250°C, 35 psi nebulizer gas and 3 kV in both positive and negative polarities for each mode. Fragmentation voltage was 380 V. For reversed phase, full scan mode collected data between 50 and 1200 mass to charge (*m/z*) and HILIC collected data between 40 and 1200 *m/z*. Two additional fragmentation experiments were performed sequentially after the full scan at 0 eV at 10 eV and 30 eV. Internal calibration mass data was also collected to maintain good accurate mass reproducibility well below 5 ppm.

3.2.3 In-house database creation

To create the in-house database 408 standards and 21 internal standards were diluted, run on both RP and HILIC columns, and data was collected in both full scan and MS/MS fragmentation at 0, 10, 20, 30, and 30 eV. Naz et al. provided a list of compounds in the supplemental data of the publisher [17]. All standards were biologically relevant in urine, plasma, or cell lysate samples, but were mainly collected by the analysts to have a broad collection of analyte polarities for testing the metabolomics workflow. The 0, 10, and 30 eV were determined to give the best fragmentation coverage of the compounds in the in-house database. The parent mass was seen at 0 eV because no fragmentation energy was applied, and no fragmentation occurred unless it was in-source fragmentation. The lower molecular weight compounds fragmented well at 10 eV, and the larger molecular weight compounds fragmented well at 30 eV. All this data is captured during the same run allowing for an efficient collection of data for multiple molecular classes.

For both RP and HILIC, the compounds were grouped based on the following retention time windows 0–3 minutes, 3 to 7 minutes, and greater than 7 minutes. This helped for tracking and correlating retention time with log P of each compound (measure of hydrophobicity). The parent compound in this study could be with or without adduct or in-source product formation. Each compound contained retention time, accurate mass measurements of the parent ion (positive and negative), adduct formation (parent ion plus sodium, ammonium, potassium, or formate), in-source fragment losses (ammonium, water, carbon dioxide, and carboxylic fragment losses), and MS/MS fragmentation. The fragments used as the qualifiers were chosen either because they were the most abundant or the most unique. The HILIC database contained 194 compounds and the RP database contained 214 compounds. It is important to note that since ionization efficiencies will differ on different instruments, running this same experiment on other instruments may change the values.

All standards were prepared neat, tested for native presence in matrix, and then spiked into matrix. The sample matrices were plasma, urine and HCT 116 cell lystate. All blank matrices were spiked with internal standard solutions and extracted with varying solvent extraction/partitioning, precipitation, centrifugation, and drying and reconstituted, mixed with standard to test for matrix effects, and tested for native amounts of the standards against the in-house database. Samples were also evaluated for any unknown compounds similar in structure to those in the database.

3.2.4 Results

All matrices did not contain every compound contained in the in-house database. To test matrix interference for every compound in all matrices, some standards were spiked into the matrices to ensure all standards were in tested in every matrix. In addition, at the time this article was written, the authors used an ion ratio threshold of +/− 25% when identifying compounds in matrix when comparing to the ion ratios of the neat standards in the in-house database. This ratio is in line, if not more conservative, than the European guidelines [19] for identification requirements of +/− 30% for two product ions using both unit mass resolution (uncertainty to 0.1 Da) and accurate mass resolution (uncertainty in 0.00001 Da). While not impacting the results of this study, it is important to note that the newest 2024 version of the European Commission guidelines [20] uses ion ratios for identification with a +/− 30% deviation for unit mass resolution only and not for accurate mass resolution. The accurate mass only requires a signal to noise ratio greater than 3, the extracted ion chromatograms of parents and ions (at least two fragment ions) must overlap, and parent and fragment data must be collected in the same run [20].

Example data for Sphingosine-1-Phosphate in **Table 1** shows how the fragmentation information was combined into descriptive data for the neat standard and for the standard spiked into the plasma matrix. The product ion/precursor ion ratio adds specificity for compound identification. The specificity is especially important in matrix. The lower relative error when comparing the neat standard ratio and standard in matrix, the less matrix effect there may be. Another important note was that the accurate mass ion ratios of the neat standard were integral in correctly identifying Sphingosine-1-Phosphate in matrix. The parent accurate mass (380.2555 Da) had a positive identification for both Sphingosine-1-Phosphate and N-palmityl-phosphoethanolamine in the Human Metabolome Database (HMDB). Without the ion ratio there would be two potential identifications rather than one.

Compounds that are structurally similar have the possibility of co-eluting on a liquid chromatography column. The specificity, as seen above, can also help when

Ion ratio	Neat standard ratio	Standards in plasma ratio	Relative error (%)
product ion 1/parent ion	0.84	0.91	−7.5
product ion 2/parent ion	0.11	0.10	6.4
product ion 3/parent ion	0.08	0.07	4.2

Table 1.
Ion ratio data for Sphingosine-1-Phosphate from data-independent acquisition or all ion fragmentation.

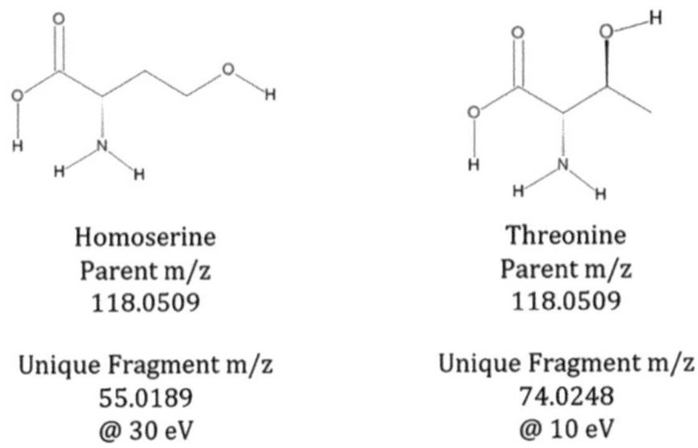

Figure 6.
All ion fragmentation data for homoserine and threonine.

co-eluting peaks have the same parent mass to charge ratio (isobaric species). The all-ion fragmentation method uses the product ions and their ratios to differentiate between isobaric compounds instead of the parent masses. In this study, both threonine and homoserine standards had the same retention time of 8.4 minutes and an accurate mass *m/z* of 118.0509. Normally these compounds could not be confidently identified or differentiated from each other, but the all-ion fragmentation protocol can form two ratios since there are two collisions energies. As seen in **Figure 6**, the structurally similar homoserine and threonine have different unique fragments at 10 eV (55.0189 *m/z*) and 30 eV (74.0248 *m/z*). These different fragments can be used to create different ion ratios and both pieces of data were collected in the same data acquisition sequence. Using the qualifier ions instead of the parent mass quantifier ions, homoserine and threonine could be resolved.

In this case study, all-ion fragmentation (data-independent acquisition) was used in combination with high-resolution mass spectrometry to create a metabolomics screening workflow in multiple matrices (plasma, urine, and cell lysate). The descriptive data was attached to each compound in the in-house database (408 compounds) using reversed phased, HILIC, and positive and negative ionization modes. All compounds had good fragmentation at either of the two different fragmentation energies and created ion ratios using the accurate mass measurements at 0, 10 and 30 eV. The workflow helps increase the accuracy of analyte identification in datasets with complex matrices.

3.3 Case study 3: HRMS and characteristic PFAS neutral fragment losses used to evaluate industrial water runoff

3.3.1 Background

Per- and polyfluoroakyl substances (PFAS) are a chemical class containing many sub-classes with unique moieties, but at a minimum PFAS compounds contain carbons and fluorides. PFAS just like hydrocarbons can also contain many other elements to create very diverse chemical moieties [21]. The structure of perfluorooctanoic acid ($C_8HF_{15}O_2$) is provided in **Figure 7** to serve as an example of a PFAS structure. Compared to petroleum-based hydrocarbons, PFAS are usually more stable, have

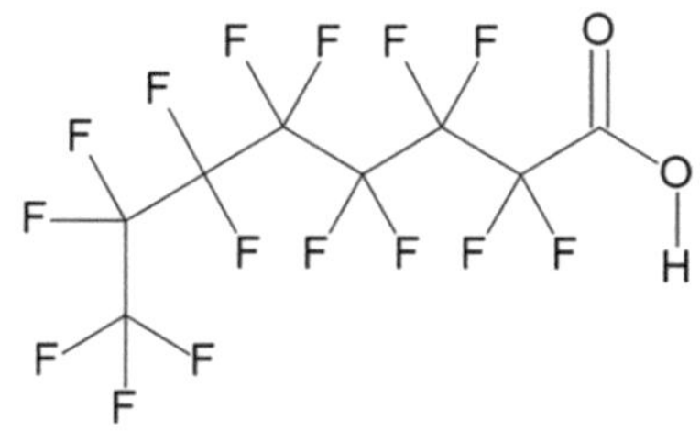

Figure 7.
Structure of perfluorooctanoic acid, a common PFAS.

higher acidity than organic compounds, and have surface altering characteristics such as water resistance, and/or oil-beading [21].

PFAS chemicals have been used since the 1940s for food packaging/contact materials, common household products, leather, clothing, fire-fighting foams, cosmetics, and agricultural chemical formulations [22]. During the production of these products or the breakdown of these products, the PFAS chemicals can wash into the river, ocean, and groundwater and can remain for many years due to their stability [23]. There is concern about the toxicity of these compounds especially if the PFAS class of compounds remain in the environment for a long time and potentially bioaccumulate in food supplies [21]. As a result, this is a need to measure the PFAS concentration and thus the toxicity of water sources.

Tang et al. have not been the first group to create an analytical method to measure PFAS levels in water, but the method they have created is a unique method to measure known and unknown PFAS in the environment [24]. This study combined accurate mass and in-source fragmentation (when a compound breaks in the electrospray ionization source) to characterize known and document unknown compounds. Neutral loss fragmentation, the loss of atoms without a charge, was the type in-source fragmentation studied. Tang et al. suggested that compared to neutral losses in collisional-induced dissociation experiments, neutral losses in the electrospray ionization sources could be more compound specific. These in-source neutral losses can produce characteristic accurate mass discrepancies between parent and the fragment ions. This can be used for structural elucidation of functional groups. Also, the ratio and isotopic distributions of the atoms such as carbon and sulfur can help with structural elucidation.

3.3.2 Instrumentation and methodology

This study used quantitative and qualitative analysis to create a hybrid non-targeted study that could quantitatively measure analytical PFAS standards and semi-quantitatively measure the new PFAS compounds. In addition, targeted experiments were done on triple quadrupole mass spectrometers to give a more accurate trace level analysis. The calibration mixture was purchased pre-mixed with 21 different PFAS compounds. In addition, four ^{13}C-labeled PFAS compounds were used as internals standards. Three different types of samples were collected for this study at a chemical manufacturing facility in China that used PFAS materials in their manufacturing process. The plant treated their wastewater with reverse osmosis, so samples were taken before and after reverse osmosis. The third sample was collected downstream of a wastewater treatment plant that was used to treat the industrial runoff and was labeled as the total effluent. This sample had gone through activated carbon filters in the wastewater treatment process.

The calibration curves were created with and without the water matrices collected at the factory to assess matrix interference with retention time and accurate mass of spiked samples. Tang et al. performed non-targeted analysis for the search of the new/unknown PFAS compounds and quasi-targeted analysis to describe their structural elucidation technique for those unknown compounds when compared to the evaluated standards [24]. In addition, non-targeted analysis used full scan and quasi-targeted analysis relied more on both data-dependent and data-independent analysis with fragmentation experiments.

Both targeted and quasi-targeted analyses used a Dionex UHPLC paired with a Thermo Q-Exactive Plus quadrupole orbitrap high-resolution mass spectrometer. The ionization source was a Thermo heated electrospray ionization source. The column used for separation was a Waters Acquity UPLC BEH C_{18} column. The LC solvents used were water with ammonium acetate and then acetonitrile. The liquid chromatography gradient was a two-step linear gradient starting at 20% organic increasing to 80%, then 95% and then re-equilibrating to 20% over 15 minutes. Ionization parameters are as follows: negative mode, −3.2 kV, capillary temperature 320°C, auxiliary gas temperature 350°C. Data was collected in full scan mode using both data-dependent and data-independent acquisition in multiple sequences.

The targeted analysis used a Waters Xevo TQD triple quadrupole tandem mass spectrometer in negative ionization mode to quantify only the commercially acquired PFAS standards, but also to help strengthen the semi-quantitation of any unknown/similar PFAS compounds. The mass spectrometer was paired with a Waters Acquity H-Class UHPLC with the liquid chromatography running solvent conditions and column being the exact same as the Q-Orbitrap method described previously. Electrospray ionization was used in negative mode, −250 V capillary voltage, 150°C source temperature, 500°C desolvation temperature, and multiple reaction monitoring (MRM) parameters for each of the different PFAS standards. Confirmation of the MRM fragmentation patterns was a way the targeted analysis could be used for structural elucidation despite usually being only for quantitation.

Non-targeted data analysis with the Q-orbitrap was quite complex since full scan with both data-dependent and data-independent acquisitions were performed. Each water sample data set of 15 minutes was divided into 0.5 seconds for full scan analysis data export. This allowed for a database to be created that included most chromatographic features which would be further classified as adducts, in-source neutral loss fragments, or in-source ionic fragments. Because a compound can have each one of these features, it is hard to assign the parent mass or not assume multiple co-eluting chromatographic peaks.

The second step for data analysis was identifying the functional groups in the PFAS as per−/polyfluorocarboxcylic acids (PFCA) or per−/polyfluoroalkanesulfonic acids (PFSA) based on their characteristic, in-source neutral losses, accurate mass isotopic differences, and in-source fragment ions. In this case, the neutral loses always occur with a loss of a separate proton based on the intrinsic properties of negative mode electrospray ionization. Since this "neutral loss" always occurs with the loss of a proton, the fragment has a negative charge, and thus can be visualized by the accurate mass Q-Orbitrap spectrometers which require ionized species or species with a mass to charge ratio (*m/z*).

As seen in **Figure 8**, the processing method assigned a PFAS as having a carboxylic acid or classified as a PFCA with the following criteria: in-source, neutral losses of 65.99172 Da (CF_2O loss) or 43.98983 Da (CO_2 loss) or a 22.00189 Da difference (CF_2O mass minus CO_2 mass) from a parent mass or any mass spectral feature. To be able

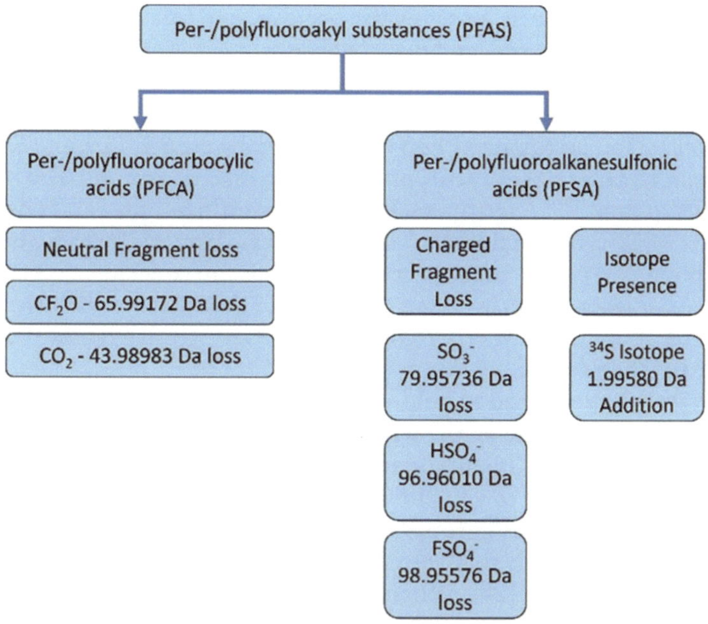

Figure 8.
PFAS classification workflow for PFCAs and PFSAs.

to classify a compound so easily based on the characteristic accurate mass difference between common, in-source neutral fragments is the strength of accurate mass spectrometry with UHPLC. As seen in **Figure 8**, the processing method assigned a PFAS as having a sulfonic acid or classified as a PFSA with the following criteria: in-source, charged fragment losses of 79.95736 Da (SO_3^- loss), 96.96010 Da (HSO_4^- loss), and/or a 98.95576 Da (FSO_4^- loss) from a parent mass or any mass spectral feature. Also, ^{34}S isotope would prominently show up around 1.99580 Da more than the parent mass which would have the ^{32}S isotope. The program also compared the ^{34}S isotope to the ^{13}C isotope for reference. Again, the criteria for PFSA determination benefit from accurate mass measurements especially for the isotopic differentiation. An important note would be without the help of the UHPLC, no features would be chromatographically resolved so this technique would not be easily employed without significant modification.

3.3.3 Results

One hundred and seventy-five PFAS were assigned a formula in the wastewater that were not included in the reference standards. Non-targeted analysis, quasi-targeted analysis, and targeted analysis were all used to assign, confirm, and measure the PFAS compounds. Eight PFCAs were measured with reference standards and ranged from 0.2 ng per mL to 23.8 ng per mL with higher values in the post reverse osmosis sample. Most of the PFAS compounds were assigned a formula through the non-targeted approach and quasi-targeted analysis was used as a complementary analysis to assign functional group designation or find additional compounds not found in non-targeted analysis. All values for these steps were semi-quantitative. Out of the 175 total PFAS compounds found, 101 could be found in all wastewater samples including pre-reverse osmosis, post-reverse osmosis, and total effluent while only 44 were unique to both the pre- and post-reverse osmosis. Among all PFAS formula identified, 55 were PFCAs and 120 were PFSAs which is 31% PFCAs and 69% PFSAs. The PFSAs and PFCAs were identified with the characteristic in-source fragments discussed earlier. Total estimated PFAS concentration were 5.3, 15.9, and 33.4 micrograms per milliliter for total effluent, pre- and post-reverse osmosis. The relative amount of PFCAs and PFSAs in the each of the total effluent, pre- and post-reverse osmosis is the same as the total PFAS. The post-reverse osmosis had a higher concentration, but it is expected because reverse osmosis can concentrate some analytes in the final post-reverse osmosis water. It is assumed that the total effluent had the lowest concentrations than the other samples due to an effective removal process used by the water treatment plant.

Additional tools were used to help with the structural elucidation workflow of the different PFAS species in the water. The main tool was database comparison of exact masses in ChemSpider or Pubchem. Also, CF_2 Kendrick masses (accurate mass normalized to CF_2 chemical groups) and adjusted Kendrick mass defect (mass defect of the CF_2 group/accumulated groups) were used to show and confirm the compounds that were related. Tang et al. included equations in their supplemental section [24]. When graphing the data according to adjusted Kendrick Mass versus the Kendrick Mass, signals that were PFAS compounds grouped together. Van Krevelen diagrams were also used to group together compounds with different amounts of hydrogens, carbons, or oxygens. Results showed the PFAS compounds in the pre- and post- reverse osmosis samples contained two groups: two to nine carbon group (highest concentration) PFAS and ten to sixteen carbon PFAS. The total effluent sample only contained the two to nine carbon PFAS group. Per-fluorinated species with zero hydrogens accounted for the highest number of 86.2% of compounds in all samples. PFAS compounds with two to four oxygens accounted for 88.8% of the samples. This data helped the analysts assign better tentative formula/structures and increase their identification confidence.

There were some PFAS compounds that could not be assigned a tentative formula or structures. Some compounds contained nitrogen and chlorine atoms. Eighty-four compounds were classified as unknown and potentially new/unreported. Additionally, three new/unreported PFSA compounds containing iodine ($C_6H_6OF_6I$-SO_3H two isomers and one isomer of $C_6H_4F_8I$-SO_3H). These three compounds made up an estimated 8.2 and 7.7% (11.9 and 23.7 ng per mL) of the total PFSAs in the pre- and post-reverse osmosis samples. This is a significant finding to investigate further in the future.

This study used non-targeted, quasi-targeted, and targeted wastewater analysis to provide quantitative and qualitative values for confirmed standards and potentially unknown compounds with no commercially available standards in environmental samples. An UHPLC Q-Orbitrap HRMS system was used to characterize in-source neutral losses and isotopic distributions for the qualitative identification process. Targeted analysis was used to measure reference PFAS standards in the wastewater samples while semi-quantitative analysis was used to measure PFAS concentrations and types in the wastewater. Roughly one hundred and seventy-five known, rarely reported, or unknown PFAS compounds were found in the wastewater samples. These results show the ability to screen and identify PFAS compounds not included in an internal library and hopefully application to additional matrices in the future.

4. Conclusions

Coupling Ultra High-Performance Liquid Chromatography with High-Resolution Mass Spectrometry adds value to many scientific investigations. This allows chromatographic separation of complex mixtures and mass spectral assignment to individual components. Because HRMS systems distinguish between small differences in *m/z*, the accurate mass values and characteristic mass defect values can be used for structure elucidation. With this descriptive data, elemental composition is possible and can confirm/assign identities and relative concentrations to known and unknown compounds. Additionally, applying certain metabolomics workflows help to simplify the analysis of the complex non-targeted analysis datasets.

Conflict of interest

The authors declare no conflict of interest.

Author details

Robert Owen Bussey III
RAI Services Company, Winston Salem, USA

*Address all correspondence to: busseyr@rjrt.com

References

[1] Arrebola-Liébanas FJ, Romero-González R, Frenich AG. HRMS: Fundamentals and basic concepts. In: Romero-González R, Frenich AG, editors. Applications in High-Resolution Mass Spectrometry. Amsterdam: Elsevier; 2017. pp. 1-14. DOI: 10.1016/B978-0-12-809464-8.00001-4

[2] de Hoffmann E, Stroobant V. Mass Spectrometry. Principles and Applications. 3rd ed. Chichester: John Wiley & Sons Ltd; 2007. p. 489

[3] Marshall AG, Blakney GT, Chen T, Kaiser KN, McKenna MA, Rodgers PR, et al. Mass resolution and mass accuracy: How much is enough? Mass Spectrometry. 2013;**2**(Spec Issue):S0009. DOI: 10.5702/massspectrometry.S0009

[4] Cooper-Shepherd DA, Wildgoose J, Kozlov B, Johnson WJ, Tyldesley-Worster R, Palmer ME, et al. Novel hybrid Quadrupole-multireflecting time-of-flight mass spectrometry system. Journal of the American Society of Mass Spectrometry. 2023;**34**:264-272. DOI: 10.1021/jasms.2c00281

[5] Kyle PB. Toxicology: GCMS. In: Nair H, Clarke W, editors. Mass Spectrometry for the Clinical Laboratory. New York: Academic Press; 2017. pp. 131-163. DOI: 10.1016/B978-0-12-800871-3.00007-9

[6] Kanawati B, Schmitt-Kopplin P. Fundamentals and Applications of Fourier Transform Mass Spectrometry. Amsterdam: Elsevier; 2019. p. 778. DOI: 10.1016/C2016-0-04234-1

[7] Shao Y, Wang Y, Yuan Y, Xie Y. A systematic review on antibiotics misuse in livestock and aquaculture and regulation implications in China. Science of the Total Environment. 2021;**798**:149205. DOI: 10.1016/j.scitotenv.2021.149205

[8] Wang X, Fu B, Yu X, Qu C, Zhang Q, Tong J. Fine mapping of growth related quantitative trait loci in Yellow River carp (Cyprinus carpio haematoperus). Aquaculture. 2018;**484**:277-285. DOI: 10.1016/j.aquaculture.2017.11.016

[9] Antimicrobial Resistance Collaborators. Global burden of bacterial antimicrobial resistance in 2019: A systematic analysis. The Lancet. 2022;**399**:629-655. DOI: 10.1016/S0140-6736(21)02724-0

[10] Xi F. The enrofloxacin pollution control from fish to environment. Marine Pollution Bulletin. 2024;**199**:115923. DOI: 10.1016/j.marpolbul.2023.115923

[11] Dai J, Wang Y, Lin H, Sun Y, Pan Y, Qiao J, et al. Residue screening and analysis of enrofloxacin and its metabolites in real aquatic products based on ultrahigh-performance liquid chromatography coupled with high-resolution mass spectrometry. Food Chemistry. 2023;**404B**:134757. DOI: 10.1016/j.foodchem.2022.134757

[12] Morales-Gutiérrez FJ, Hermo MP, Barbosa J, Barrón D. High-resolution mass spectrometry applied to the identification of transformation products of quinolones from stability studies and new metabolites of enrofloxacin in chicken muscle tissues. Journal of Pharmaceutical and Biomedical Analysis. 2014;**92**:165-176. DOI: 10.1016/j.jpba.2014.01.014

[13] Xiao C, Han Y, Liu Y, Zhang J, Hu C. Relationship between fluoroquinolone structure and neurotoxicity revealed

by zebrafish neurobehavior. Chemical Research in Toxicology. 2018;**4**(30):238-250. DOI: 10.1021/acs.chemrestox.7b00300

[14] Brennan L. Metabolomics in nutrition research-a powerful window into nutritional metabolism. Essays Biochemistry. 2016;**60**(5):451-458. DOI: 10.1042/EBC20160029

[15] Johnson CH, Ivanisevic J, Siuzdak G. Metabolomics: Beyond biomarkers and towards mechanisms. Nature Reviews. Molecular Cell Biology. 2016;**17**(7):451-459. DOI: 10.1038/nrm.2016.25

[16] Smith CA, O'Maille G, Want EJ, Qin C, Trauger SA, Brandon TR, et al. METLIN: A metabolite mass spectral database. Therapeutic Drug Monitoring. 2005;**27**(6):747-751. DOI: 10.1097/01.ftd.0000179845.53213.39

[17] Naz S, Gallart-Ayala H, Reinke SN, Mathon C, Blankley R, Chaleckis R, et al. Development of a liquid chromatography-high-resolution mass spectrometry metabolomics method with high specificity for metabolite identification using all ion fragmentation acquisition. Analytical Chemistry. 2017;**89**(15):7933-7942. DOI: 10.1021/acs.analchem.7b00925

[18] Sumner LW, Amberg A, Barrett D, Beale MH, Beger R, Daykin CA, et al. Proposed minimum reporting standards for chemical analysis chemical analysis working group (CAWG) metabolomics standards initiative (MSI). Metabolomics. 2007;**3**(3):211-221. DOI: 10.1007/s11306-007-0082-2

[19] EU Reference Laboratory (EURL) for Residues of Pesticides. Analytical Quality Control and Method Validation Procedures for Pesticide Residues Analysis in Food and Feed. Solarex. Document SANTE/11945/2015 2016. Available from: https://eurl-pesticides.eu/library/docs/allcrl/AqcGuidance_SANTE_2015_11945.pdf [Accessed: 27 June 2024]

[20] EU Reference Laboratory (EURL) for Residues of Pesticides. Analytical Quality Control and Method Validation Procedures for Pesticide Residues Analysis in Food and Feed. SANTE/11312/2021 (V2). 2024. Available from: https://food.ec.europa.eu/system/files/2023-11/pesticides_mrl_guidelines_wrkdoc_2021-11312.pdf [Accessed: 27 June 2024]

[21] Wang Z, DeWitt JC, Higgins CP, Cousins IT. A never-ending story of per- and Polyfluoroalkyl substances (PFASs)? Environmental Science and Technology. 2017;**51**:2508-2518. DOI: 10.1021/acs.est.6b04806

[22] KEMI Swedish Chemicals Agency. Report 7/15: Occurrence and Use of Highly Fluorinated Substances and Alternatives. 2020. Available from: https://www.kemi.se/download/18.6df1d3df171c243fb23a98ea/1591454109137/report-7-15-occurrence-and-use-of-highly-fluorinated-substances-and-alternatives.pdf. [Accessed: 28 June 2024]

[23] Heydebreck F, Tang J, Xie Z, Ebinghaus R. Alternative and legacy perfluoroalkyl substances: Differences between European and Chinese river/estuary systems. Environmental Science and Technology. 2015;**49**:8386-8395. DOI: 10.1021/acs.est.5b01648

[24] Tang C, Liang Y, Wang K, Liao J, Zeng Y, Luo X, et al. Comprehensive characterization of per- and polyfluoroalkyl substances in wastewater by liquid chromatography-mass spectrometry and screening algorithms. Npj Clean Water. 2023;**6**:1-6. DOI: 10.1038/s41545-023-00220-6

Chapter 2

Optimizing Ultra-High-Performance Liquid Chromatography (UHPLC) Methods for Rapid Analysis of Complex Biological Samples

Ariaya Hymete, Feruza Ahmed, Ayenew Ashenef and Tekeste Abebe

Abstract

Ultra-High-performance liquid chromatography (UHPLC) offers significant advantages for analyzing complex biological samples, including faster analysis time, improved resolution, and enhanced sensitivity. However, the inherent complexity of biological matrices and diverse analyte properties pose challenges. This chapter explores strategies for optimizing UHPLC methods to achieve rapid and reliable analysis. Key parameters like column selection, mobile phase composition, and sample preparation steps are covered in this chapter. Recent advancements in UHPLC technology, such as sub-2 μm particle columns and UHPLC systems, are addressed. The chapter review provides a valuable resource for researchers seeking to leverage UHPLC for the efficient analysis of complex biological samples.

Keywords: UHPLC, chromatography, optimization, complex samples, biological samples

1. Introduction

Ultra-High-performance liquid chromatography (UHPLC) has emerged as a powerful analytical tool in biological research due to its superior speed, resolution, and sensitivity compared to traditional high-performance liquid chromatography (HPLC) [1]. Shortened analysis time is particularly valuable in biological research, where time-sensitive samples or those requiring rapid analysis for downstream applications are frequently encountered [2]. Thus, it enables researchers to rapidly analyze complex biological samples, which are often characterized by a wide range of biomolecules with diverse physicochemical properties. Furthermore, UHPLC delivers superior resolution, enabling researchers to differentiate between closely related analytes that might co-elute using traditional HPLC methods [3]. This is crucial for unraveling the complex chemical landscape of biological samples, which often harbor a diverse range of biomolecules with subtle physicochemical variations.

Moreover, UHPLC provides heightened sensitivity, allowing for the detection and quantification of analytes present at low concentrations within biological matrices [4]. It is essential for uncovering low-concentration biomarkers or metabolites that might hold significant biological implication.

However, the inherent complexity of biological samples presents challenges for UHPLC analysis. Biologicals are often an intricate mixture of proteins, lipids, carbohydrates, and other biomolecules, which can create a phenomenon known as the "*matrix effect*" [5]. "Matrix effect" is the phrase commonly used to characterize issues that arise when analyzing complex biological samples. Endogenous (such as proteins, lipids, or metabolites of the target analyte) or exogenous (all substances added during sample processing and analysis) components are typically responsible for it [6]. Matrix effects can interfere with analyte detection by suppressing or enhancing ionization, leading to inaccurate quantification. Additionally, the diverse physicochemical properties of analytes within a single biological sample can make chromatographic separation difficult [3]. For instance, highly hydrophobic analytes might elute very close to each other, hindering accurate identification and quantification.

This chapter review delves into strategies for optimizing UHPLC methods to overcome challenges and achieve rapid and reliable analysis of complex biological samples.

1.1 Challenges of analyzing complex biological samples with UHPLC

There's probably no universal agreement on what constitutes a complex sample. One aspect of complexity may be seen by many analysts as the combination of chemicals from completely distinct substance families (e.g., proteins, carbohydrates, plant secondary compounds, and foreign substances). The presence of some analytes at trace levels and others with concentrations in the high (g/kg) range could be the other factor. When an analyte appears as a substantial element in one sample but as a trace compound in another, the situation becomes even more complicated [7].

The intent of quantitative bioanalytical method is to offer a precise and accurate estimation of the concentration of a target analyte in complicated biological samples (fluids or tissues), which is typically a drug, metabolite, or biomarker. The demand for biological sample analysis requires multidisciplinary work that includes drug metabolism, clinical analysis, pharmacokinetic investigations, therapeutic drug monitoring (TDM), and forensic analysis.

Over the past few years, the development of bioanalytical methods has become increasingly difficult due to extremely strict requirements regarding method reliability, sensitivity, analysis speed, and sample throughput [8]. Additionally, the high salt content of some biological samples, like urine, can negatively impact UHPLC performance by reducing chromatographic resolution and column life [5].

While UHPLC offers significant advantages for biological analysis, its effectiveness can be hampered by the inherent complexity of biological samples [9]. This is due to samples being an intricate tapestry of biomolecules, including proteins, lipids, carbohydrates, and secondary metabolites, each with distinct physicochemical properties [10]. A number of issues, including the existence and intensity of endogenous and exogenous interferences (phospholipids and dosing vehicles), analyte instability during storage and/or in an ion source, and the requirement for extremely rapid, extremely sensitive, and reliable methods, pose significant challenges for the bioanalytical scientist [8]. The two main challenges for UHPLC analysis are matrix effects and the associated ion suppression or enhancement and analyte diversity that causes co-elution [10].

1.2 Matrix effects and ion suppression

The intricate composition of biological samples can lead to a phenomenon known as the "matrix effect" [11]. During the ionization process, often within the mass spectrometer hyphenated UHPLC [2], co-eluting biomolecules can interact with the analyte of interest, influencing its ability to become ionized [12]. The interactions can either suppress or enhance the analyte's signal, causing erroneous quantification. It is particularly problematic for low-concentration analytes, potentially masking their presence entirely [13].

It has been demonstrated that matrix effects depend on the ionization techniques utilized for the liquid chromatography-mass spectrometry (LC-MS) method, which in the majority of contemporary liquid chromatography–tandem mass spectrometry (LC-MS/MS) experiments are either atmospheric pressure chemical ionization (APCI) or electrospray ionization (ESI) [14]. Whether analyte and co-eluting mixture components outcompete one another during the ionization process depends on their respective chemical structures and concentration levels [15]. For instance, because ESI has a significant preference toward surfactants that enrich at the droplet surface during the liquid–/gas-phase ion transfer, it is especially sensitive to co-eluting phospholipids [16]. In other words, analyte ions that are trapped inside droplets may not be able to go out due to the presence of phospholipids on their surface. On the other side, because there is little compound competition to enter the gas phase of the mass spectrometer, APCI is frequently less impacted by suppression effects. Thus, ion suppression is more likely to affect the ESI than APCI [17]. Regardless, matrix effects are experienced by APCI in multicomponent data. Ion suppression effects will almost always be present in any LC-MS/MS assay because biofluids contain a large number of endogenous molecules, frequently at high stages, with potentially very high basicity and surface activities [6]. Theoretically, ionization effects can happen in the gas phase or the solution phase [17].

The severity of matrix effects depends on the specific biological sample and the analyte of interest. For instance, analyzing hydrophobic drugs in blood plasma can be challenging due to the presence of abundant proteins that can compete for ionization sites, leading to signal suppression [18, 19]. Similarly, analyzing endogenous metabolites in tissue extracts can be hindered by the presence of lipids, which can cause ion suppression effects [20].

Generally speaking, using stable isotope-labeled internal standards (SIL-ISs) for each analyte appears to be the most dependable method of accounting for matrix effects. However, it is also not a foolproof solution. Serious matrix suppression may result in a significant reduction in sensitivity, which is undesirable for samples with extremely low analyte concentrations. Additionally, a number of studies revealed an unexpected behavior for deuterium-labeled SIL-ISs, that the analyte to internal standard (IS) peak area ratio significantly altered when a different batch of plasma was used or when the analytical conditions were slightly altered. Due to a small variation in retention under specific circumstances, the ionization of the analyte was impacted differently than those of its isotopically labeled equivalents, which resulted in significant quantification errors [21–23]. The deuterium isotope effect causes this phenomenon because substituting a hydrogen atom for a deuterium results in a little change in lipophilicity. Due to its reduced lipophilicity compared to the parent molecule, the deuterated IS elutes a little bit early in the chromatogram [24].

1.3 Analyte diversity and co-elution

Complex biological samples harbor a vast array of biomolecules with diverse physicochemical properties, including polarity, size, and charge [25]. This diversity can make achieving optimal chromatographic separation using UHPLC a challenge [26]. Analytes with similar physicochemical properties may elute from the UHPLC column at the same time, resulting in overlapping peaks on the chromatogram [15]. The co-elution hinders accurate identification and quantification of individual analytes, especially for closely related species or those present at low concentrations.

Imagine, for example, analyzing a bacterial culture extract containing various fatty acid derivatives. Two fatty acids with very similar chain lengths and functionalities might elute very close together on the UHPLC column, appearing as a single broad peak. This makes it difficult to distinguish and quantify each individual fatty acid within the sample [25].

2. Method optimization parameters for complex biological samples analyses by UHPLC

Optimizing UHPLC methods for complex biological samples necessitates meticulous attention to various parameters. By strategically fine-tuning the required parameters, researchers can achieve efficient separation of analytes, minimize matrix effects, and significantly improve the overall sensitivity and accuracy of their analyses [9]. This section delves into key UHPLC method optimization parameters, exploring their influence and offering practical considerations for method development.

2.1 Sample preparation

Isolation, cleanup, and/or preconcentration of target analytes from intricate biological matrices are the main goals of the sample preparation stage [8].

Effective sample preparation is critical for successful UHPLC analysis of complex biological samples. The standard method in the analysis of blood plasma and serum is protein precipitation (PP), which is based on an increase in intermolecular hydrophobic interactions and decrease in protein hydration followed by the addition of a suitable precipitation agent [27]. PP is the simplest sample preparation technique, and it is also a way whereby, proteins can be eliminated using ultrafiltration cut-off membranes or denatured with acids or heat [28].

Organic solvents for protein precipitation can also be used. According to studies, methanol extracts are less pure than tetrahydrofuran or ethanol extracts and include 40% more phospholipids than acetonitrile [29]. For instance, numerous articles highlight the application of UHPLC-MS/MS for anticoagulant detection and quantification. Acetonitrile or methanol was utilized as a sample cleaning in the majority of these procedures, which involved plasma and protein precipitation. Phospholipids, which can accumulate on the analytical column and contaminate the mass spectrometer source, leading to matrix effects (MEs) and decreased sensitivity, are not eliminated by this type of precipitation [6, 30].

Another study was conducted for the purpose of identifying beta-blockers in blood, a sample preparation technique combining the simultaneous use of protein precipitation and salting-out homogeneous liquid-liquid extraction was devised [27].

Solid-phase extraction [31, 32] and derivatization can also help remove interferences, concentrate analytes, and improve the detection limits of the method [13]. On the other hand, highly selective sample preparation techniques are convenient to minimize matrix effects and ionization alterations that could occur with ESI-MS detection [8].

Pretreatment typically entails a liquid extraction stage, followed by processes to eliminate or minimize co-extracted matrix elements, such as lipids, proteins, and carbohydrates [33–35].

A chiral UHPLC-MS method for the analysis of cysteine enantiomers (Cys) in biological samples indicated that assessing native Cys in biological samples continues to be difficult and is often done after the -SH groups have stabilized during the preparation of the sample. This is explained by the fact that Cys are unstable in extracellular fluid and quickly autoxidize to cystine [36, 37].

A method validation and analysis of forensic samples indicated that an enzymatic hydrolysis step was suggested to be included in the sample preparation step to detect phase II metabolites from opioids and synthetic cannabinoids [38].

By integrating optimized sample preparation protocols with UHPLC methods, researchers can achieve faster analysis times, higher sensitivity, and improved data quality for complex sample matrices [21, 39].

2.2 Column selection

The UHPLC column serves as the foundation for achieving optimal separation of analytes within a complex biological sample. The choice of column plays a critical role in this process, influencing factors like selectivity, resolution, and peak shape [25].

A study on a high-throughput UHPLC/MS/MS-based metabolic profiling using a vacuum- jacketed column indicated that the reduced column length and increased gradient steepness (column volumes/minute) resulted in sharper peaks. Conventional systems face issues with the sharpness of these smaller peak volumes being adversely impacted with intrinsic dispersion and thermal dispersion due to frictional heating. The short 2.1 × 30 mm vacuum-jacketed column (VJC), constructed using vacuum-jacketed stainless-steel tubing and a porous stationary phase, was designed to minimize these effects. The column's inlet and outlet temperatures were maintained at different levels to minimize thermal dispersion, and the column was connected to the MS source using a fused silica capillary minimizing extra column dispersion [40].

A comparison of a novel ultra-performance liquid chromatographic method for determination of retinol and α-tocopherol in human serum with conventional HPLC using monolithic and particulate column showed that an Acquity UPLC BEH (Ethylene Bridged Hybrid) C18 column used for analysis of protein-precipitated rat plasma maintained initial peak capacity and selectivity for over 2200 injections [41].

Here are key considerations for column selection:

a. *Stationary phase type:* The stationary phase refers to the material coated on the inner surface of the UHPLC column. Common stationary phases include:

i. *Reversed-phase (RP):* It is the most widely used type for biological sample analysis. It is the most widely used type of stationary phase for biological sample analysis. Analytes are separated based on their hydrophobicity, with more hydrophobic analytes interacting more strongly with the RP stationary phase, leading to longer retention times [42]. By measuring the quantity of double bonds, the length of their non-polar side chains, and the class of lipids they belong to, reversed-phase liquid chromatography can be used to separate lipids [43, 44]. Thermo Fisher Scientific Hypercarb, Waters XTerra C18, Phenomenex Kinetex C18, Agilent Zorbax C8 and SB-CN (StableBond Packing is made by chemically bonding a sterically protected cyanide (CN) stationary phase), and Phenomenex Kinetex pentafluorophenyl-bonded phase (PFP) were among the reversed-phase materials evaluated for the analytes' chromatographic separation. The PFP column produced the best findings because it was able to separate the highly polar N-acetyl-P-aminophenol (APAP) metabolites from other, less polar analytes and resolve these metabolites [28]. The reversed-phase techniques that are most commonly used apply a stationary phase with carbon chain lengths of C8 or C18 [45, 46].

ii. *Hydrophilic interaction chromatography (HILIC):* HILIC offers balanced selectivity to RP for separating highly polar analytes that might not retain well on traditional RP columns [47]. The unique benefits of hydrophilic interaction liquid chromatography (HILIC) include no biases toward different glycopeptides, good compatibility with MS, high throughput, and low cost [48]. HILIC performs better than alternative enrichment or separation techniques. Acetonitrile (CH_3CN) is typically the best option for the aqueous-organic mobile phase in HILIC, which may separate polar and hydrophilic solutes using a polar substance [49].

iii. *Ion-exchange:* It is useful for separating analytes based on their charge. This can be particularly valuable for biological samples containing charged molecules like proteins and peptides [50].

b. *Particle size:* Smaller particle sizes (sub-2 µm) offer superior resolution but may require higher pressure to achieve optimal flow rates. Conversely, larger particle sizes (e.g., 5 µm) are suitable for lower pressure operation but may compromise resolution [51].

c. *Pore size:* The pore size of the stationary phase allows for the diffusion of analytes into the interior of the packing material. Smaller pore sizes can enhance selectivity but may also increase backpressure. Selecting the appropriate pore size depends on the size and complexity of the analytes of interest [52].

3. Mobile phase composition

When it comes to guiding the analytes of a complex biological sample through the chromatographic separation process, the mobile phase type and composition is essential. Analytes' interactions with the stationary phase are greatly influenced by its composition, which ultimately determines their retention times and separation. Researchers can attain the best possible separation and a well-balanced performance for all of the analytes present by carefully adjusting the composition of the mobile phase.

3.1 Organic solvents: setting the stage for polarity

The organic solvent, typically mixed with water, forms the primary component of the mobile phase. Common choices include acetonitrile (ACN), methanol (MeOH), and isopropanol (IPA) [42]. The selection of the organic solvent has a profound impact on the overall polarity of the mobile phase. This, in turn, dictates the strength of analyte interactions with the stationary phase. More polar analytes will interact more strongly with a polar mobile phase, leading to longer retention times. Conversely, less polar analytes will elute faster with a more non-polar mobile phase [26].

- *Acetonitrile (ACN):* It is the most widely used organic solvent in UHPLC due to its low viscosity, high volatility, and miscibility with water across a broad range [9]. It offers a good balance between polarity and elution strength, making it suitable for separating a wide range of analytes in complex biological samples [51].

- *Methanol (MeOH):* A more polar solvent compared to ACN, it offers increased selectivity for separating highly polar analytes. However, its higher viscosity can lead to higher backpressure in UHPLC systems [42]. In order to design and validate a chiral UHPLC-MS method for the detection of cysteine enantiomers in biological samples, MeOH/methyl cyanide (MeCN)/water (H2O) (49/49/2, v/v/v) with 50 mM ammonium formate ($HCOONH_4$) and 50 mM formic acid were utilized as the optimized mobile phase composition. The utilized mobile phases were degassed before usage [37].

- *Isopropanol (IPA):* It offers a unique combination of properties, being less polar than MeOH but more polar than ACN. This can be advantageous for separating analytes with intermediate polarities [25].

The choice of organic solvent, or even a combination of them, is often dictated by the specific analytes of interest within the complex biological sample [19].

3.2 Water additives: fine-tuning analyte ionization

While water is a common component of the mobile phase, it's not merely a passive carrier. The addition of water modifiers, such as formic acid (HCOOH), acetic acid (CH_3COOH), or ammonium formate, plays a crucial role in further fine-tuning the mobile phase composition [50]. Development and validation of a UHPLC-HRMS-QTOF (high-resolution mass spectrometry-quadrupole time-of-flight) method for the detection of 132 new psychoactive substances and synthetic opioids had a mobile phase that consisted of water (A) and acetonitrile (B), both with formic acid 5 mM [53].

These additives can:

- *Adjust mobile phase pH:* Water additives can act as weak acids or bases, influencing the mobile phase pH. This can be particularly important for separating charged analytes. For instance, using formic acid as an additive can lower the mobile phase pH, which can improve the peak shape and separation of basic analytes [18].

- *Enhance analyte ionization:* Certain additives, like ammonium formate, can act as ion-pairing agents, facilitating the ionization of analytes in the ESI source often coupled with UHPLC [20]. This can improve sensitivity for certain analytes, particularly those that are poorly ionized on their own. The polarity of the substances affects their ionization and, consequently, their signal acquisition using mass spectrometry, making the LC-MS analysis of the simultaneous detection of the endogenous anabolic androgenic steroids (EAAS) in human serum difficult. While their conjugated metabolites of glucuronides and sulfates have enough polarity to ionize in the negative ESI, unconjugated EAAS have low polarity and consequently poor ionization. In the negative ESI mode of LC-MS, buffer solutions with a pH of 4.5 and higher or bases like ammonium hydroxide are frequently used as mobile phase additions. Although weakly basic low polarity unconjugated EAAS would normally favor ionization in the positive ESI mode, this can be improved by adding acids like formic acid [31].

The selection and concentration of water additives are crucial for optimizing analyte separation and sensitivity in UHPLC analysis of complex biological samples.

3.3 Gradient elution: a dynamic approach for diverse analytes

A powerful technique for separating a varied range of analytes with diverse polarities present in complex biological samples is gradient elution [10]. Unlike isocratic elution where the mobile phase composition remains the same, gradient elution dynamically changes the mobile phase composition over time. This allows for the elution of a broader range of analytes with varying polarities within a single UHPLC run [9].

The gradient profile, which defines the starting and ending time solvent composition, and the rate of change between them (gradient slope), is carefully optimized to achieve the desired separation [42]. A typical gradient elution for biological samples might start with a high percentage of water (polar) and a low percentage of organic solvent (non-polar) in the mobile phase. As the run time progresses, the percentage of organic solvents gradually increases using a defined gradient slope (e.g., linear, convex). This creates a less polar environment that elutes the more hydrophobic analytes. This approach allows for the separation of both polar and non-polar analytes within a single UHPLC run, overcoming the limitations of isocratic elution for complex biological samples [26].

On a study done for the analysis of prescription drugs, synthetic cathinones, phenethylamines, synthetic cannabinoids, amphetamines, and other psychoactive substances, gradient elution was performed with a constant flow rate of 0.5 mL/min and a column oven temperature of 40°C, utilizing the following gradient: 0–0.5 minutes: 1% B; 4 minutes: 30% B; 7 minutes: 60% B; 9 minutes: 70% B; 10–12 minutes: 99% B; and 12.1 minutes: 1% B. Subsequently, B was held at 1% for 2 minutes for column equilibration. The total run time equates to 14.1 minutes [38].

A simultaneous quantification of acetaminophen along with structurally related compounds used a gradient elution at a flow rate of 300 μl/min. The total run time of analysis was 19 minutes. The column temperature was maintained at 40°C during separation [28].

Optimizing the mobile phase composition is an iterative process that often involves evaluating different combinations of organic solvents, water additives, and gradient profiles. By carefully considering the specific analytes of interest within the complex biological sample and their physicochemical properties, researchers can create a mobile phase composition that coordinates a successful separation performance.

3.4 Practical considerations for mobile phase optimization

Building upon the understanding of organic solvents, water additives, and gradient elution, this section can offer practical guidance for researchers on how to approach mobile phase optimization:

- *Starting point selection:* Researchers can leverage existing literature or established methods for similar biological matrices as a starting point. This provides a foundational mobile phase composition to build upon [25].

- *Systematic adjustments:* Fine-tuning can involve making small, incremental changes to one parameter at a time (e.g., organic solvent percentage, water additive concentration, gradient slope) while monitoring the impact on analyte separation. This systematic approach allows researchers to identify the effect of each parameter on the chromatography [42].

- *Monitoring analyte behavior:* For example: peptides have multiple charge states, diluting the ion signal. Nonspecific binding (NSB) is another challenge, causing non-linearity, poor sensitivity, carry-over, and peak tailing. Solutions like organic solvent-rich, pH-modified, or adsorption competitors can help reduce NSB by interfering with hydrophobic interactions, preventing ionic or hydrophobic interactions, and blocking interaction sites at sample preparation or LC-system surfaces [54].

  Throughout the optimization process, it's crucial to monitor the separation and peak shape of the analytes of interest. Techniques like peak resolution calculations and peak symmetry assessments can be used to evaluate the effectiveness of mobile phase adjustments [9].

- *Utilizing chemometrics:* Advanced software tools employing chemometrics can be valuable for researchers. These tools can analyze chromatographic data and suggest potential mobile phase modifications to achieve optimal separation [50, 55, 56].

4. Addressing matrix effects

Suppression of ionization: The predominant matrix effect discovered was ion suppression. Ion suppression is a significant element that can impact a mass detector's quantitative performance. Recent ESI investigations on biological extracts have demonstrated that the presence of nonvolatile or less volatile solutes alters the spray droplet solution's characteristics, which is the primary source of ion suppression. The efficiency of droplet formation is altered by certain nonvolatile substances (such as salts, ion-pairing agents, endogenous chemicals, and medications/metabolites) [17]. By carefully selecting mobile phase components and optimizing the pH, researchers can minimize competition between analytes and matrix components for ionization sites in the ESI source [18].

Up to 29.4% ion suppression for 6-monoacetylmorphine (6-MAM) and norfentanyl was reported [57]. Eighty-nine for phencyclidine (PCP) in serum [58], up to 39% for synthetic cannabinoids in whole blood [59], and up to 29.3 and 39.4% for amphetamine and morphine in urine was seen [60].

A determination of new psychoactive substances and other drugs in postmortem blood and urine by UHPLC–MS/MS indicated that ion suppression was seen in urine for 6-MAM and 1-(5-fluoropentyl)-3-(1-naphthoyl) indole (AM 2201) at a medium level (36.1 and 35.1%, respectively), and amphetamine, codeine, methamphetamine, and morphine (28.8–37.1%) at the lowest concentration level. In blood, significant ion suppression (28.0–37.2%) was found at the lowest level for 6-MAM, benzoylecgonine, codeine, methamphetamine, and morphine. Given that 12 matrix-analyte combinations showed substantial matrix effects (> ± 25%), matrix-matched calibration curve was done for quantification purpose [38].

To achieve ion suppression, the easiest way is to change the chromatography's ionization source from ESI to APCI. The samples can be readily reanalyzed under the new circumstances with only a little source modification needed [61].

Besides, a study conducted on the analysis of cysteine enantiomers in biological samples showed that, to overcome co-elution problems expected in biological samples the specificity of a single quadrupole mass spectrometry detector was used [37].

5. Salting effects

High salt content presence in some biological samples can be addressed by incorporating volatile salts (e.g., ammonium acetate) into the mobile phase. These salts can improve peak shape and chromatographic resolution. The use of ammonium acetate should be controlled, with a maximum concentration of 0.1 M. This method is frequently utilized in place of phosphate buffers, which ought never to be applied in conjunction with LC-MS interfaces [5].

6. Conclusion

Ultra-high-performance liquid chromatography (UHPLC) has become a cornerstone technique for unraveling the intricate tapestry of complex biological samples. Its ability lies in its use to separate a vast array of analytes with unparalleled resolution and speed.

Nonetheless, the complex nature of biological samples poses a considerable obstacle to its exceptional capacity to isolate a wide range of analytes. The matrix effect and co-elution are the most frequent problems encountered. In addition, maximizing UHPLC's analytical potential hinges on a crucial step—mobile phase optimization. The transformative benefits of a well-optimized mobile phase include enhanced separation of diverse analytes, improved sensitivity for low-abundance species, minimized peak distortion, and ultimately, high-quality data for downstream analysis techniques [62]. Furthermore, the emergence of sub-2 μm particle stationary phases with superior resolving power opens doors to even more complex separations within biological samples [63]. These advanced stationary phases, coupled with optimized mobile phase compositions, will allow researchers to differentiate between closely related analytes and achieve a deeper coverage of the biological sample's intricate biochemical landscape.

Conflict of interest

There is no conflict of interest in writing this chapter.

Author details

Ariaya Hymete[1*], Feruza Ahmed[1], Ayenew Ashenef[1,2] and Tekeste Abebe[3]

1 Department of Pharmaceutical Chemistry and Pharmacognosy, School of Pharmacy, College of Health Sciences, Addis Ababa University, Addis Ababa, Ethiopia

2 Center for Innovative Drug Development and Therapeutic Trials for Africa (CDT-Africa), College of Health Sciences, Addis Ababa University, Addis Ababa, Ethiopia

3 Pharmacy School, College of Health Sciences and Medicine, Wolaita Sodo University, Wolaita Sodo, Ethiopia

*Address all correspondence to: ariaya.hymete@aau.edu.et

References

[1] Gregorova P, Sipari NH, Peter Sarin L. Broad-range RNA modification analysis of complex biological samples using rapid C18-UPLC-MS. RNA Biology. 2020;**18**(10):1382-1389

[2] Kurbanoglu S, Karsavurdan O, Ozkan SA. Recent advances on drug analyses using ultra performance liquid chromatographic techniques and their application to the biological samples. Current Analytical Chemistry. 2019;**15**(3):277-293

[3] Rathod RH, Chaudhari SR, Patil AS, Shirkhedkar AA. Ultra-high performance liquid chromatography-MS/MS (UHPLC-MS/MS) in practice: Analysis of drugs and pharmaceutical formulations. Future Journal of Pharmaceutical Sciences. 31 Oct 2019;**5**(1):1-26

[4] Smith J, White K, Johnson P, Wilson M. Advances in Ultraperformance liquid chromatography for rapid analysis of complex biological samples. Journal of Separation Science. 2018;**41**(13):2598-2612

[5] Nguyen TD, Ong ES. Overcoming challenges in modern separation science for metabolomics. Trends in Analytical Chemistry. 2018;**102**:138-150

[6] Bylda C, Thiele R, Kobold U, Volmer DA. Recent advances in sample preparation techniques to overcome difficulties encountered during quantitative analysis of small molecules from biofluids using LC-MS/MS. The Analyst. 2014;**139**(10):2265

[7] Kaufmann A. Combining UHPLC and high-resolution MS: A viable approach for the analysis of complex samples? TrAC Trends in Analytical Chemistry. 2014;**63**:113-128

[8] Nováková L. Challenges in the development of bioanalytical liquid chromatography–mass spectrometry method with emphasis on fast analysis. Journal of Chromatography. 2013;**1292**:25-37

[9] Han X, He W, Guo Z, Xu H, Jiang H, Wang Y. Recent advances in ultra-performance liquid chromatography for the analysis of complex biological samples in environmental and food safety. Trends in Analytical Chemistry. 2020;**123**:115724

[10] Vorkas PA, Isaac G, Anwar MA, Davies AH, Want EJ, Nicholson JK, et al. Untargeted UPLC-MS profiling pipeline to expand tissue metabolome coverage: Application to cardiovascular disease. Analytical Chemistry. 2015;**87**(8):4184-4193

[11] Cappiello A, Famiglini G, Palma P, Trufelli H. Matrix effects in liquid chromatography-mass spectrometry. Journal of Liquid Chromatography & Related Technologies. 2010;**33**(9-12):1067-1081. Available from: http://quimica.udea.edu.co/~carlopez/cromatohplc/matrix-effects-in-lc-ms-2010.pdf

[12] Cimino G, Contin M, Favaro G, Franch J, Magno C, Robotti E, et al. Matrix effects in quantitative LC-MS analysis of new psychoactive substances in human plasma. Journal of Chromatography. 2008;**1204**(1-2):142-149

[13] Chen H, Wang J, Kong W, Xu L, Li Y, Liu Y. Matrix effect in UHPLC-MS/MS analysis of mycotoxins in cereals and effective minimization approaches. Food Chemistry. 2018;**240**:1174-1182

[14] Matuszewski BK, Constanzer ML, Chavez-Eng CM. Strategies for

the assessment of matrix effect in quantitative bioanalytical methods based on HPLC–MS/MS. Analytical Chemistry. 2003;**75**(13):3019-3030. DOI: 10.1021/ac020361s

[15] Volmer DA. Ion Suppression: A Major Concern in Mass Chromatography Spectrometry. 2021. Available from: https://www.chromatographyonline.com/view/ion-suppression-major-concern-mass-spectrometry [Accessed: August 20, 2024]

[16] Janusch F, Kalthoff L, Hamscher G, Mohring SAI. Evaluation and subsequent minimization of matrix effects caused by phospholipids in LC–MS analysis of biological samples. Bioanalysis. 2013;**5**(17):2101-2114

[17] Annesley TM. Ion suppression in mass spectrometry. Clinical Chemistry. 2003;**49**(7):1041-1044

[18] Xu Y, Liang Y, He Y. Matrix effect in quantitative analysis of hydrophobic drugs in human plasma by UHPLC-MS/MS. Separation and Purification Technology. 2019;**212**:669-677

[19] De Baere S, Ochieng PE, Kemboi DC, Scippo M-L, Okoth S, Lindahl JF, et al. Development of high-throughput sample preparation procedures for the quantitative determination of aflatoxins in biological matrices of chickens and cattle using UHPLC-MS/MS. Toxins. 2023;**15**(37):1-24

[20] Annesley TM. Ion suppression in metabolite profiling using liquid chromatography–mass spectrometry. Electrophoresis. 2018;**39**(12-13):1610-1620

[21] Cazorla-Reyes R, Romero-González R, Frenich AG, Rodríguez Maresca MA, Martínez Vidal JL. Simultaneous analysis of antibiotics in biological samples by ultra-high performance liquid chromatography–tandem mass spectrometry. Journal of Pharmaceutical and Biomedical Analysis. 2014;**89**:203-212. Available from: https://reader.elsevier.com/reader/sd/pii/S0731708513005086?token=FF3C0BC366BFBC505D32AA453C627EBB3AC7CA8993C946F800E1A5020E6EE1D6559D4807759616F18EC29B6DD332D2FB&originRegion=eu-west-1&originCreation=20230414072122

[22] NiklasLindegardh AA, White NJ, Nicholas. Development and validation of a liquid chromatographic-tandem mass spectrometric method for determination of piperaquine in plasma. Biomed Research International. 2008;**862**(1-2):227-236

[23] Jemal M, Schuster A, Whigan DB. Liquid chromatography/tandem mass spectrometry methods for quantitation of mevalonic acid in human plasma and urine: Method validation, demonstration of using a surrogate analyte, and demonstration of unacceptable matrix effect in spite of use of a stable isotope analog internal standard. Rapid Communications in Mass Spectrometry. 2003;**17**(15):1723-1734

[24] Berg T, Strand DH. 13C labelled internal standards—A solution to minimize ion suppression effects in liquid chromatography–tandem mass spectrometry analyses of drugs in biological samples. Journal of Chromatography A. 2011;**1218**(52):9366-9374

[25] Talebi M, Jalali-Khanlar H, Valizadeh H. Optimization of UHPLC methodforthesimultaneousdetermination of multiple classes of pesticides in various food matrices. Food Chemistry. 2019;**272**:77-86

[26] Zhao J, Liu Y, Li J. Recent developments in ultra-performance liquid chromatography coupled with mass spectrometry for the analysis of biological samples. Analytica Chimica Acta. 2019;**1075**:1-20

[27] Yıldırım S, Fikarová K, Pilařová V, Nováková L, Solich P, Horstkotte B. Lab-in-syringe automated protein precipitation and salting-out homogenous liquid-liquid extraction coupled online to UHPLC-MS/MS for the determination of beta-blockers in serum. Analytica Chimica Acta. 2023;**2023**(1251):340966. Available from: https://www.sciencedirect.com/science/article/abs/pii/S0003267023001873

[28] Bylda C, Thiele R, Kobold U, Volmer DA. Simultaneous quantification of acetaminophen and structurally related compounds in human serum and plasma. Drug Testing and Analysis. Aug 2013;**6**(10):1045-1053

[29] Chambers E, Wagrowski-Diehl DM, Lu Z, Mazzeo JR. Systematic and comprehensive strategy for reducing matrix effects in LC/MS/MS analyses. Journal of Chromatography. 2007;**852**(1-2):22-34. Available from: https://www.waters.com/webassets/cms/library/docs/720002135en.pdf

[30] Jemal M, Ouyang Z, Xia YQ. Systematic LC-MS/MS bioanalytical method development that incorporates plasma phospholipids risk avoidance, usage of incurred sample and well thought-out chromatography. Biomedical Chromatography. 2010;**24**(1):2-19

[31] Elmongy H, Masquelier M, Ericsson M. Development and validation of a UHPLC-HRMS method for the simultaneous determination of the endogenous anabolic androgenic steroids in human serum. Journal of Chromatography A. 2020;**1613**:460686

[32] Baldelli S, Cattaneo D, Pignatelli P, Perrone V, Pastori D, Radice S, et al. Validation of an LC–MS/MS method for the simultaneous quantification of dabigatran, rivaroxaban and apixaban in human plasma. Bioanalysis. 2016;**8**(4):275-283

[33] Xie LJ, Chen M, Ying YB. Development of methods for determination of aflatoxins. Critical Reviews in Food Science and Nutrition. 2016;**56**:2642-2664

[34] Escriva L, Manyes L, Font G, Berrada H. Mycotoxin analysis of human urine by LC-MS/MS: A comparative extraction study. Toxins. 2017;**9**:330

[35] Flores-Flores ME, Gonzalez-Penas E. An LC-MS/MS method for multi-mycotoxin quantification in cow milk. Food Chemistry. 2017;**218**:378-385

[36] Jones DP, Carlson JL, Mody V, Cai J, Lynn M, Sternberg P. Redox state of glutathione in human plasma. Free Radical Biology and Medicine. 2000;**28**(4):625-635

[37] Pucciarini L, González-Ruiz V, Zangari J, Martinou J-C, Natalini B, Sardella R, et al. Development and validation of a chiral UHPLC-MS method for the analysis of cysteine enantiomers in biological samples. Journal of Pharmaceutical and Biomedical Analysis. 2020;**2020**(177):112841-112841

[38] Ferrari Júnior E, Caldas ED. Determination of new psychoactive substances and other drugs in postmortem blood and urine by UHPLC–MS/MS: Method validation and analysis of forensic samples. Forensic Toxicology. 2022;**40**:88-101. DOI: 10.1007/s11419-021-00600-y

[39] Zhang K, Banerjee K. A review: Sample preparation and chromatographic

technologies for detection of aflatoxins in foods. Toxins. 2020;**12**:539

[40] Plumb RS, McDonald T, Rainville PD, Hill J, Gethings LA, Johnson KA, et al. High-throughput UHPLC/MS/MS-based metabolic profiling using a vacuum jacketed column. Analytical Chemistry. 2021;**93**(30):10644-10652

[41] Citová I, Havlíková L, Urbánek L, Solichová D, Nováková L, Solich P. Comparison of a novel ultra-performance liquid chromatographic method for determination of retinol and α-tocopherol in human serum with conventional HPLC using monolithic and particulate columns. Analytical and Bioanalytical Chemistry. 2007;**388**(3):675-681

[42] Wu N, He W, Li L. Recent advances in reversed-phase high-performance liquid chromatography columns for metabolomics. Journal of Chromatography A. 2016;**1459**:8-19

[43] Carrasco-Pancorbo A, Navas-Iglesias N, Cuadros-Rodríguez L. From lipid analysis towards lipidomics, a new challenge for the analytical chemistry of the 21st century. Part I: Modern lipid analysis. TrAC Trends in Analytical Chemistry. 2009;**28**(3):263-278

[44] Teo CC, Chong WPK, Tan E, Basri NB, Low ZJ, Ho YS. Advances in sample preparation and analytical techniques for lipidomics study of clinical samples. TrAC Trends in Analytical Chemistry. 2015;**66**(1):1-18

[45] Cajka T, Fiehn O. Comprehensive analysis of lipids in biological systems by liquid chromatography-mass spectrometry. TrAC Trends in Analytical Chemistry. 2014;**2014**(61):192-206. Available from: https://www.ncbi.nlm.nih.gov/pmc/articles/PMC4187118/

[46] Rustam YH, Reid GE. Analytical challenges and recent advances in mass spectrometry based lipidomics. Analytical Chemistry. 2017;**90**(1):374-397

[47] Lai N, Xu W, Zhou M. Recent advances in hydrophilic interaction liquid chromatography columns for metabolomics. Electrophoresis. 2019;**40**(1):102-114

[48] Ahn YH, Kim JY, Yoo JS. Quantitative mass spectrometric analysis of glycoproteins combined with enrichment methods. Mass Spectrometry Reviews. 2015;**34**:148-165. DOI: 10.1002/mas.21428

[49] Qing G, Yan J, He X, Li X, Liang X. Recent advances in hydrophilic interaction liquid interaction chromatography materials for glycopeptide enrichment and glycan separation. TrAC Trends in Analytical Chemistry. Jun 2020;**124**:115-126

[50] Amat E, Soga T, Wakabayashi T. Ion exchange chromatography for metabolomics. Current Opinion in Biotechnology. 2018;**55**:1-8

[51] Guillarme D, Heinisch S, Ruegger M. The role of particle size in sub-2-μm particulate stationary phases for ultra-performance liquid chromatography. Journal of Chromatography. 2010;**1217**(16):2477-2490

[52] Dolan JW. Peak capacity: Definition, variation with column characteristics and its use in high-performance liquid chromatography method optimization. Journal of Chromatography A. 2010;**1216**(9):6366-6376

[53] Massano M, Incardona C, Gerace E, Negri P, Alladio E, Salomone A, et al. Development and validation of a UHPLC-HRMS-QTOF method for

the detection of 132 new psychoactive substances and synthetic opioids, including fentanyl, in dried blood spots. Talanta. 2022;**241**:123265

[54] Bongaerts J, De Bundel D, Smolders I, Mangelings D, Heyden YV, Van Eeckhaut A. Improving the LC-MS/MS analysis of neuromedin U-8 and neuromedin S by minimizing their adsorption behavior and optimizing UHPLC and MS parameters. Journal of Pharmaceutical and Biomedical Analysis. 2023;**228**:115306-115306. Available from: https://pubmed.ncbi.nlm.nih.gov/36868028/

[55] Alrobaian M, Panda SS, Almalki WH, Afzal O, Kazmi I, Alossaimi MA, et al. Development and validation of chemometrics-assisted green UPLC-MS/MS bioanalytical method for simultaneous estimation of capecitabine and lapatinib in rat plasma. Journal of Chromatographic Science. 2021;**60**(6):559-570

[56] Bos TS, Boelrijk J, Molenaar SRA, van't Veer B, Niezen LE, van Herwerden D, et al. Chemometric strategies for fully automated interpretive method development in liquid chromatography. Analytical Chemistry. 2022;**94**(46):16060-16068

[57] Orfanidis A, Gika HG, Theodoridis G, Mastrogianni O, Raikos N. A UHPLC-MS-MS method for the determination of 84 drugs of abuse and pharmaceuticals in blood. Journal of Analytical Toxicology. 2020;**45**(1):28-43

[58] Lehmann S, Kieliba T, Beike J, Thevis M, Mercer-Chalmers-Bender K. Determination of 74 new psychoactive substances in serum using automated in-line solid-phase extraction-liquid chromatography-tandem mass spectrometry. Journal of Chromatography B. 2017;**1064**(1):124-138

[59] Odoardi S, Fisichella M, Romolo FS, Strano-Rossi S. High-throughput screening for new psychoactive substances (NPS) in whole blood by DLLME extraction and UHPLC–MS/MS analysis. Journal of Chromatography B. 2015;**1000**(1):57-68

[60] Yang CA, Liu HC, Lin DL, Liu RH, Hsieh YZ, Wu SP. Simultaneous quantitation of methamphetamine, ketamine, opiates and their metabolites in urine by SPE and LC–MS-MS. Journal of Analytical Toxicology. 2017;**41**(8):679-687

[61] Morgan DG, Olah TV. Ion suppression and cannulation locking solutions. Bioanalysis. 2009;**1**(3):523-528

[62] Jandera P. Stationary phases for high-performance liquid chromatography. Journal of Chromatography A. 2018;**1568**:1-25

[63] Wang Y, Gao X, Li Y, Mao Y, Chen Y. Recent advances in sub-2 μm particle stationary phases for high-performance liquid chromatography. Journal of Chromatography. 2020;**2020**(1627):461342

Chapter 3

Theory and Recent Applications of Nano-Liquid Chromatography

Esayas Tesfaye, Ayenew Ashenef, Ariaya Hymete and Tadios Niguss

Abstract

The trend toward instrument miniaturization in recent years has made it possible to develop new and sophisticated analytical techniques, such as nano-liquid chromatography (nano-LC). This has made it possible to improve the sensitivity and resolution of chromatography. Nano HPLC is essential for both qualitative and quantitative analytical methods. The growing interest in nano-LC methods has led to the development of several fascinating and inventive applications. This chapter will cover the theoretical aspects of the nano-LC method and its current practical uses in the analysis of pharmaceutical and biological molecules. Furthermore, the future prospects regarding the development of nano-LC techniques will be examined.

Keywords: nano-liquid chromatography, injection system, analytical performance, pharmaceutical, biomedical

1. Introduction

It is not novel to see systems becoming smaller. Miniaturization is the current trend. According to statistics published in the literature in the 1978–1990s, miniaturization was introduced by researchers in the field of liquid chromatography (LC) approximately 30 years ago. Early research, for example, by the groups of Novotny and Knox, examined theoretical as well as practical aspects of lowering the column internal diameter (i.d.) that is used in chromatography [1–4]. More recently, biological applications in the proteomics research field have been the main driving force behind the expansion of miniaturized LC devices. The developments in micro-column liquid chromatography will be covered in this chapter, with particular emphasis on column development, instrumentation, analyte detection, and application areas.

The analytical discipline is the primary application for miniature techniques including nano-LC, which use stationary phases (SPs) housed within capillaries (fused silica). Typical example of this, capillary electrophoresis (CE), one of these methods with several modes, which has been extensively researched and used [5–11]. However, only nano-liquid chromatography, or nano-LC, will be covered in this chapter.

2. Theory and principle of nano-liquid chromatography

2.1 Theoretical aspects of nano-LC

The foundation principle of a tiny chromatographic system is the same as that of normal chromatography. The injected analytes may dilute in the column during the chromatographic process, changing the effectiveness of the separation process:

$$D = \frac{Co}{Cmax} = \frac{\Pi d_c^{\,2}\varepsilon(1+k)\sqrt{2LH\Pi}}{Vinj} \tag{1}$$

The expression for this dilution event, known as chromatographic dilution (D), is as follows: where C_o represents the starting concentration, C_{max} denotes the analyte's final concentration during the chromatography process, dc denotes the inner diameter of the column, ε denotes column porosity, L denotes column length, $4V_{inj}$ denotes sample injection volume, k and H are the chromatographic parameters retention factor and plate height, respectively. The formula states that as the square of the column diameter decreases, D reduces correspondingly. A considerable drop in D-values is made possible by lower i.d. nano-LC as opposed to traditional HPLC. Therefore, less chromatographic dilution is required with a smaller chromatographic system, improving the separation's mass detectability [12]. Miniaturizing LC systems should theoretically be very advantageous for liquid phase separations. However, because they lead to decreased separation efficiency, a few real-world separation challenges need to be taken into account.

2.2 Analytical instrumentation of nano-LC systems

2.2.1 Pump

In order to facilitate nanoscale gradient elution, pumps for nano-LC should have stable, repeatable nanoflow rates throughout separations. For the Nano-LC, there are two primary systems available: split pumps and splitless pumps, the latter of which are offered for sale and are commercially available. Split systems distinguish high flow rates (mL/min) from traditional HPLC pumps by using a flow restrictor between the pump and a smaller column. With a straightforward nanoflow restrictor design, these devices can utilize conventional HPLC pumps [13]. Split systems, however, have the potential to result in inconsistent split ratios, inconsistent nanoflow, and inconsistent separation. It is quite challenging to achieve reproducible gradient elution, particularly when using DIY (Do it yourself) split devices. This manner of elution may be limited by differences in backpressure caused by different mixed solvent viscosities [14]. Splitless systems are currently commonly employed in nano-LC. These technologies provide extremely reproducible nanoflow rates while eliminating solvent loss. Split systems are inferior to syringe pumps with a single limited-volume reservoir; however, continuous-flow pumps, which resemble traditional piston pumps with two pistons per channel, are now the most widely used type. Both isocratic and gradient nanoflow elutions can be performed using continuous-flow pumps, and adjusting to the required nanoflow rate is simple.

2.2.2 Injection

Another crucial factor in chromatography that requires careful consideration is the volume of the injected sample. Inadequate injection volume can lead to major issues and worsen the analytes' separation by broadening the column band. As a result, the sample must be injected into any chromatographic system as narrowly as feasible; the issue becomes more apparent when the column's i.d. decreases. The maximum sample volume that can be injected into a column depends on several parameters such as the length of the column (L), the particle diameter (dp), the column diameter (dc), and the retention factor (k):

$$V_{max} = 0.36\sqrt{Ldpdc}\,(K+1) \tag{2}$$

As a result, specific, commercially accessible valves with internal loop capacities of 40–60 nL are utilized. As an alternative, loops with larger volumes are used; however, in this instance, the time is controlled by injecting only a portion of the sample solution [15]. Therefore, an LC injector with minor changes would be required for a laboratory-made nano-LC system (see **Figure 1**) in addition to a split pump system and a UV-visible detector. Lastly, automated sample injection techniques are used in current nano-LC instrumentation, where volumes—typically μL—are split into nL.

2.2.3 Nano columns

While 10 μm i.d. columns are also an option, 75 μm i.d. nano-LC columns are the most commonly utilized. In nano-LC separations, this i.d. column offers a reasonable balance between robustness, loadability, and detectability.

Since there were no commercially available columns in the early days of nano-LC, researchers had to make their own. Thanks to effective packing techniques, the first nano-LC columns were made commercially available in the 1990s. The selection of columns has expanded significantly over the past 20 years, and nano-LC columns are also offered. Research laboratories utilize self-packed nano-LC columns frequently for cost-effective purposes on a periodic basis. There are several variants and formats of nano-LC columns, monoliths, open tubular columns, and the pillar array format. Most applications are done with packed columns, while the monolith and open tubular columns are still less established as routine tools. The two types of nano-LC columns are explained below.

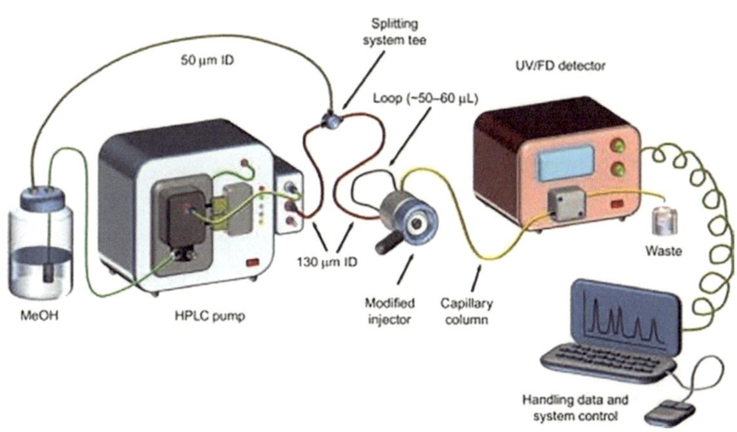

Figure 1.
Schematic diagram of assembled nano-LC instrumentation.

a. Packed columns

Fused silicon capillaries covered with polyimide are utilized to create the packed columns used in the nano-LC columns. In addition to their current flexibility, strong mechanical resistance, and range of internal dimensions, nanocolumns can also be made of titanium and stainless steel tubes. They can be filled with a monolithic bed of silica-based particles, packed with silica-based particles, or, much less frequently, wall-coated with the proper inorganic or organic materials. Particle sizes between 3 and 5 μm are most frequently used in packed nanocolumns. Particle-filled tiny i.d. columns are challenging to prepare, nevertheless.

b. Monolithic columns

In biospecific analysis, biocompatible compounds present an intriguing alternative to conventional synthesis methods for processing monoliths. Because the stationary phase is fixed to the column wall in this sort of column, a porous (silica or polymer) structure forms throughout the column, obviating the requirement for frits. Single rods of organic or inorganic material that can form in the capillary column are known as monolithic stationary phases. Monolithic columns do not require frits, and their high porosity permits faster mobile phase flow rates, which shortens the separation time. Modern nano-LC columns with the separation power needed for these complicated proteomics samples are now commercially available in lengths as long as 50 cm.

2.2.4 Analyte detection

The kinds of detection used in HPLC separations are also used in nano-LC. Nano-LC frequently uses diode array detection (DAD) due to its low cost, broad applicability, and usage of online detection. Nevertheless, when using on-column detection, detectability is restricted because of the nano-column's small path length. Longer light routes are provided by specially designed detecting cells, which solve this problem [16]. Although they are employed in nano-LC detection, laser-induced fluorescence [17] and inductively coupled plasma MS [18] are not reliable enough to be used in routine analysis. Applications in biomedicine and pharmacology typically need for universal detection techniques with high detectability, like MS detection. For MS coupling across different nanospray interfaces, particularly electrospray ionization (ESI), which only needs a tiny quantity of eluent from the LC column to be successful, the nanoflow from the column (often 100–500 nL/min) is sufficient [19].

3. Application of nano-liquid chromatography

Numerous studies and reviews on miniaturized nano-LC, covering theoretical issues as well as practical applications, support these methods as effective tools that are also more environmentally friendly than the existing separation techniques, particularly when combined with MS.

3.1 Pharmaceutical application

Pharmaceutical analysis is a significant subject with a wide range of applications in various domains, including drug quality control, pharmacokinetic research, and examination of the chiral purity and quality of pharmaceutical formulations. For chiral analysis, nano-LC and capillary LC are favored methods of quality control; however HPLC is still the method of choice for industrial quality control. Actually, a lot of research is being done on novel chiral stationary phases (CSPs), which are frequently synthesized in little quantities or with low yields yet sufficient to build a capillary column.

One of the applications of Nano-LC on pharmaceutical analysis is the enantiomeric separation of amlodipine and its two chiral impurities by nano-liquid chromatography and capillary elecrochromatography using a chiral stationary phase based on cellulose tris. In this work, a novel polysaccharide-based chiral stationary phase (CSP), cellulose tris (4-chloro-3 methylph enylcarbamate), also called Sepapak 4 has been evaluated for the chiral separation of amlodipine, and its two impurities. Capillary columns of 100 μm id packed with the chiral stationary phase were used for both nano-liquid chromatography (nano-LC) and capillary elecrochromatography (CEC) experiments. To fully resolve the mixture of six enantiomers, parameters such as buffer pH and concentration sample injection have been then investigated. A mixture of ACN/water (90/10, v/v) containing 5 mM ammonium borate buffer pH 9.0 enabled the complete separation of the three couples of enantiomers in less than 30 minutes [20].

Another study shows a new nano-liquid chromatographic method for b-blocker enantiomers' separation. This method uses a capillary column packed with silica particles, which were chemically modified with vancomycin. On-column focusing allowed the inject of relatively high sample volumes (1500 nL), increasing method sensitivity. The studied racemic compounds, namely atenolol, propranolol, oxprenolol, and metoprolol, were dissolved in methanol and injected for chromatographic separation. The effect of injected sample volume was studied in the range of 50–2100 nL. Under optimal experimental conditions, LODs and LOQs (LOD and LOQ for each alprenolol enantiomer) were 9.0 and 15.6 ng/mL, respectively. Calibration curves were linear in the studied range (9–250 ng/mL). The optimized method was applied to the analysis of a human plasma sample spiked with racemic alprenolol [21].

The other application focuses on the development of a microfluidic confocal fluorescence detection system for the hyphenation of nano-LC to online biochemical assays. Here, they worked toward minimization of sample and reagent consumption by coupling nano-LC online to a light emitting diode (LED) based capillary confocal fluorescence detection system capable of online BCD with low-flow rates. In this fluorescence detection system, a capillary with an extended light path (bubble cell) was used as a detection cell in order to enhance sensitivity. The current setup uses 50 nL as injection volume with a carrier flow rate of 400 nL/min. Finally, coupling of the detection system to gradient reversed-phase nano-LC allowed analysis of mixtures in order to identify the bioactive compounds present by injecting 10 nL of each mixture [22]. Some applications of nano-LC in pharmaceutical analysis are explained in **Table 1**.

3.2 Biomedical application

It is essential to quickly and highly accurately identify molecules of biological interest. Recent advancements in analytical instrumentation and sample preparation techniques have facilitated biological investigations aimed at identifying these remarkable compounds. Nowadays, medicinal and veterinary medications, doping management, disease diagnosis, and the quantitative identification of biomarkers and proteomes are the main application sectors for nano-LC analysis; the latter is mainly

Analytes	Column	Detector	Mobile phase	Comments	References
Racemic mixtures of 50 multiclass pharmaceutical	20 cm × 150 μm i.d. S.P.: CSP- single-walled carbon nanotubes (SWCNTs) in polymer monolithic backbone	UV λ = 219–270 nm	Isocratic (300 nL/min) CH_3OH/acetonitrile/ water/Trifluoroacetic acid 2-propanol/ CH_3OH *n*-hexane/2-propanol	Single-walled carbon nanotubes (SWCNTs) encapsulated into different polymer-based monolithic backbones for monomer, cross linker and porogen solvent	[20]
Racemic mixtures of warfarin, naproxen	18/25 cm × 75 μm i.d. S.P.: silica, S.S-Whelk-O1–2.5-CSP (2.5 μm, 120 Å)	UV λ = 214nmMS (Orbitrap)	Isocratic (300 nL/min) Warfarin: n-hexane/ CH_2Cl_2/CH_3OH Naproxen: CH_3OH / $NH_4CH_3CO_2$	Single polymeric organic monolithic outlet frit. RP separation of no chiral hydrophobic mixture	[23]
Racemic mixtures of 8 Non-steroidal anti-inflammatory drugs(NSAIDs)	10 cm × 100 μm i.d. S.P.: C18-monolithic silica or C18 (3 μm, 110 Å)	UV λ = 200 nm	Isocratic (−) Methanol/water/ Hydroxypropyl-β-Cyclodextrin/ CH_3COONa, pH	Chiral mobile phase additives (CMPAs) chiral analysis. Comparison between monolithic silica and C18 silica particles (3 μm, 110 Å)	[24]
Racemic mixtures of temazepam, thalidomide, warfarin, etozoline	25 cm × 75 μm i.d. S.P.: Chiral stationary phase(CSP)-silica based cellulose tris(4-chloro-3-methylphenylcarbamate) coated (10% w/w) (i) 3 μm native silica particles (ii) core-shell silica (2.8 μm)	UV λ = 205 nm	Isocratic (150 nL/min) Acetonitrile(CAN)/ water/ $NH_4CH_3CO_2$, pH 4.5	Comparison between porous and core shell silica based CSP	[25]
7 Sympathomimetic Drugs	25 cm × 100 μm i.d. S.P: cross-linked diol hydrophilic interaction liquid chromatography (HILIC) phase (μm, 195 Å)	UV λ = 205 nm	Isocratic (300 nL/min) Acetonitrile(ACN)/ water/ NH_4HCO_2, pH 3	Comparison between polar SPs (i.e., cyano-, diol-, aminopropylsilica and hydrophilic interaction liquid chromatography (HILIC) phase	[21]

Analytes	Column	Detector	Mobile phase	Comments	References
Amlodipine and two Impurities	30 cm × 100 μm i.d. S.P.: Chiral stationary phases(CSP)silica based on cellulose tris (4-chloro-3-methylphenylcarbamate)	UV λ = 206 nm	Isocratic (100 nL/min) Acetonitrile(ACN)/ water/ NH_4-borate, pH 10	Chiral analysis of Impurities	[14]
8 Steroids	5 cm × 50 μm i.d. S.P.: C18 hydrate (1.8 μm, 100 Å)	UV λ = 200–254 nm	Steroids: Isocratic (380 nL/min) Acetonitrile(ACN)/ water/ NSAIDs: Isocratic (300 nL/min) Acetonitrile(ACN)/ water/	Comparison between different i.d. columns (50, 75 and 100 μm)	[22]
7 Phenoxy acid herbicides	20 cm × 100 μm i.d. S.P.: poly MQD-coHEMA-co-EDMA	UV λ = 210 nm	Isocratric (20 μL/min) Acetonitrile (ACN)/ water/ $NH_4CH_3CO_2$, pH 5.	Chiral analysis	[26]
Acetaminophen, paracetamol, aspirin metabolites	Monolithic column (375μm x 50 μm i.d. with a length of 28 cm)	UV λ = 205 nm	0.05 mL/min Acetonitrile (ACN)/ water/ Tris buffer (20 mmol/L, pH 8.5)	Hydrophilic and strong anion exchange interaction	[27]

Table 1.
Applications of nano-LC in pharmaceutical analysis.

due to the extremely small sample size needed. HPLC-based methods overcome the classical problems of protein analysis, such as gel electrophoresis and immunoanalysis, which are both limited by multiple steps before analyses. The introduction of nano-LC coupled to MS and MS-MS has increased the need for quick and reliable identification methodologies due to the complexity of the proteome diversity. With the use of a comprehensive identification database, they have made it possible to precisely determine the amino acid sequences from proteins or peptides. Nevertheless, because a combination of two or more identification methodologies yields a wealth of information about protein sequencing and peptide mapping, classical approaches are still employed with nano-LC-MS. Using nano-LC–MS-MS, proteomic investigations have been carried out on synovial fluid from rheumatoid arthritis patients. Destructive articular illnesses including osteoarthritis and rheumatoid arthritis are marked by inflammatory abnormalities when defense cells gradually break down the cartilage tissues. Peptides associated with both articular disorders and additional peptides unique to each were found in one investigation. Understanding the synovial fluid proteome was crucial for identifying protein fractions that served as biomarkers and facilitated effective clinical management of patient care [28].

One biomedical application was to identify epithelial and stromal proteins that exhibit up- or down-regulation in keratoconus (KC) versus normal human corneas. Because previous proteomic studies utilized whole human corneas or epithelium alone, thereby diluting the specificity of the proteome of each tissue, they selectively analyzed the epithelium and stromal proteins. Individual preparations of epithelial and stromal proteins from KC and age-matched normal corneas were analyzed by two independent methods, that is, a shotgun proteomic using a Nano-Electrospray Ionization Liquid Chromatography Tandem Mass Spectrometry [Nano-ESI-LC-MS $(MS)^2$] and two-dimensional-difference gel electrophoresis (2D-DIGE) coupled with mass spectrometric methods. The label-free Nano-ESI-LC-MS $(MS)^2$ method identified 104 epithelial and 44 stromal proteins from both normal and KC corneas, and also quantified relative changes in levels of selected proteins, in both the tissues using spectral counts in a proteomic dataset [29].

A similar relative analysis of stroma by this method also showed upregulation of aldehyde dehydrogenase 3A1 (ALDH3A1), keratin 12, apolipoprotein A-IV precursor, haptoglobin precursor, prolipoprotein and lipoprotein Gln in keratoconus corneas. Together, the results suggested that the nano-ESI-LC-MS (MS) 2 method was superior to the 2D-DIGE method as it identified a greater number of proteins with altered levels in KC corneas. Further, the epithelial and stromal structural proteins of KC corneas exhibited altered levels compared to normal corneas, suggesting that they are affected due to structural remodeling during KC development and progression. Additionally, because several epithelial and stromal enzymes exhibited up- or down-regulation in the KC corneas relative to normal corneas, the two layers of KC corneas were under metabolic stress to adjust their remodeling [29].

In another biomedical application of nano-LC, the therapeutic monoclonal antibodies (mAbs) constitute a group of highly effective agents for treating various refractory diseases. Nonetheless, it is challenging to achieve selective and accurate quantification of mAb in pharmaceutical matrices, which is required by PK studies. In this study, they employed a suite of technical advances to overcome these difficulties, which include: (i) a nano-LC/SRM-MS approach to achieve high analytical sensitivity, (ii) a high-resolution nano-LC/LTQ/Orbitrap for confident identification of candidate peptides, (iii) an on-the-fly orthogonal array optimization (OAO) method for the high-throughput, accurate and reproducible optimization for numerous candidate peptides in a single

Matrix	Proteome	Nano-LC conditions	References
Human corneas	Proteins from healthy and keratoconus corneas	5 µL extracted tissue injected; C18 (5 µm, 100 µm i.d. ×11 cm); Acetonitrile(ACN) gradient, 500 nL/min; ion trap analyzer	[29]
Candida albicans	Surface proteins from cell wall	10^8 digested cells injected; C18 (3 µm, 75 µm i.d *15 cm); CAN gradient, 300 nL/min, TOF analyzer after UV detection	[31]
Marine shells	Proteins from layers of the shell	1 µL 10^8 digested sample injected;C18 (3 µm, 75 µm i.d *15 cm); ACN gradient, 50 nL/min, TOF analyzer	[32]
Preclinical mouse modes	Monoclonal antibodies from mouse plasma	2 µL 108 treated sample injected; C18 (5 µm, 75 µm i.d *25 cm); ACN gradient, 250 nL/min, no trap analyzer	[30]

Table 2.
Recent proteomic and peptide analyses using nano-LC techniques coupled to MS detection.

LC/MS run without using synthesized peptides, (iv) a comprehensive evaluation of stability of candidates in matrix using the optimized SRM parameters, (v) the use of two unique SP for quantification of one mAb to gauge possible degradation/modification in biological system and thus enhancing data reliability (e.g., rejection of data if the deviation between the two SP is greater than 25%), and (vi) the utilization of purified target protein as the calibrator to eliminate the risk of severe negative biases that could occur when a synthesized peptide is used as calibrator [30].

To show a proof of concept, this strategy is applied in the quantification of cT84.66, a chimeric, anti-CEA antibody, in preclinical mouse models. The strategy employed in this study can be easily adapted to the sensitive and accurate analysis of other mAb and therapeutic proteins [30]. Some applications of nano-LC on biomedical analysis including the proteome and its nano-LC conditions are explained in **Table 2**.

3.3 Environmental analysis

The use of nano-LC for the analysis of compounds of environmental interest has not been so widely extended up to now, although HPLC is one of the major techniques for the analysis of pollutants and their metabolites. In fact, very few works deal with the application of nano-LC in environmental analysis [33].

MS detection was coupled through a nanospray interface, obtaining good intra- and interday precision and complete separation of zeralanol (ZAL) and E2 isomers in less than 20 minutes. The nano-LC-MS method was applied for the detection of the target compounds in mineral waters, extracting the analytes by SPE with commercial zearalenone (ZEN)-molecular imprinted polymer (MIP) cartridges, and obtaining LODs. The method was compared with CEC-MS, but less satisfactory limits of detection (LOD) and limit of quantification (LOQ) results were achieved due to the lower injection with respect to nano-LC and the MS signal suppression caused by the presence of the buffer in the MP [33].

Also, regarding monolithic columns, Rahayu and coworkers *in situ* prepared a 300 µm i.d. monolithic column attaching polyethylene glycol (PEG) groups into a glycidyl methacrylate monolith polymer for the separation of inorganic anions (IO3-, BrO3-, NO3-, Br-, and NO3-). It constituted a very simple and fast method to analyze seawater and public drinking water, and a simple filtration was needed before injection [34].

3.4 Food analysis

The applications of miniaturized LC techniques to the analysis of foodstuffs are scarce with reference to other fields. The analysis of anthocyanins in commercial red fruit juices was assessed by nano-LC, and therefore the method was compared with conventional HPLC by Fanali and coworkers. A C18 capillary column of 100 μm i.d. and MS was employed within the first case, while a 2.1 mm i.d. narrow-bore C18 column and diode array detector (DAD) was utilized in the second. Both methods were fully validated to achieve higher sensitivity with HPLC. However, nano-LC offered good quantitative leads to a shorter analysis time. The column, full of fully porous particles, allowed the separation of all studied compounds in but 7 minutes and gave better results with respect to the column containing partially porous particles. On the opposite hand, the nano-LC system provided higher separation efficiency and worse peak symmetry and determination than conventional C18 columns by HPLC. The developed nano-LC method was applied to wine pomace samples and represented a suitable system for the analysis of anthocyanin dyes [15].

Besides the control of food quality and nutraceutical analysis, food safety has experienced an excellent rising concern. Food products reach the buyer through human handling and action, which generally results in the introduction of exogenous molecules that could modify and accelerate food deterioration or even endanger human health. In recent years, nano-LC and CLC have been applied for the determination of harmful compounds in food matrices such as antibiotics/drugs, pesticides, mycotoxins, and endocrine-disrupting compounds (EDCs) like phthalates [35].

3.5 Future prospects

Nano-LC techniques are emerging as an accessible, robust, and easy-to-use platform. The hyphenation techniques of nano-LC with Mass spectrometry are offering the sensitivity, selectivity, and high resolution demanded in the analyses of highly regulated biopharmaceuticals. Thus, future analytical works in complex biological samples will rely heavily on nano-LC techniques.

4. Conclusions

As this chapter has shown, nano-LC is a cutting-edge separation method with a lot of promise, particularly in the analytical scale arena. It has been used in a variety of industries, including biomedical interest, agro-chemical, food, pharmaceutical, and environmental analyses techniques and samples. Pharmaceutical and biological applications require methodologies that are sensitive enough to identify and measure. Chemicals with biological significance are present in trace amounts. Hence, the used procedures, as demonstrated by nano-LC–MS and nano-LC–MS-MS hyphenations, must have outstanding detectability and undeniable identification, particularly for these low-concentration compounds. As an adjunct to immunoassays and electrophoresis in the analysis of biological molecules, nano-LC may soon establish a dominant position. In the near future, however, nano-LC has the potential to reach a consolidated position in the analysis of biological molecules as a complement to electrophoresis and immunoassays.

DOI: http://dx.doi.org/10.5772/intechopen.1006657

Author details

Esayas Tesfaye[1*], Ayenew Ashenef[2,3], Ariaya Hymete[2] and Tadios Niguss[4]

1 Department of Pharmacy, College of Health and Medical Science, Dilla University, Dilla, Ethiopia

2 Department of Pharmaceutical Chemistry, School of Pharmacy, College of Health Science, Addis Ababa University, Addis Ababa, Ethiopia

3 Center for Innovative Drug Development and Therapeutic Trials for Africa (CDT-Africa), College of Health Science, Addis Ababa University, Addis Ababa, Ethiopia

4 Department of Research and Training, Eka Kotebe General Hospital, Addis Ababa, Ethiopia

*Address all correspondence to: esa.tesfaye7@gmail.com

References

[1] Tsuda T, Novotny M. Packed microcapillary columns in high performance liquid chromatography. Analytical Chemistry. 1978;**50**:271-275

[2] Tsuda T, Novotny M. Band-broadening phenomena in microcapillary tubes under the conditions of liquid chromatography. Analytical Chemistry. 1978;**50**:632-634

[3] Knox JH, Gilbert MT. Kinetic optimization of straight open-tubular liquid chromatography. Journal of Chromatography. 1979;**186**:405-418

[4] Tsuda T, Hibi K, Nakanlshl T, Takeuchi T, Ishii D. Studies of open-tubular micro capillary liquid chromatography: II. Chemically bonded octadecylsilane stationary phase. Journal of Chromatography. 1978;**158**:227-232

[5] Fanali S. Identification of chiral drug isomers by capillary electrophoresis. Journal of Chromatography. A. 1996;**735**:77-121

[6] Chankvetadze B. Enantioseparations by using capillary electrophoretic techniques: The story of 20 and a few more years. Journal of Chromatography. A. 2007;**1168**:45-70

[7] Jáč P, Scriba GKE. Recent advances in electrodriven enantioseparations. Journal of Separation Science. 2013;**36**:52-74

[8] Calcerrada M, González-Herráez M, García-Ruiz C. Recent advances in capillary electrophoresis instrumentation for the analysis of explosives. TrAC Trends in Analytical Chemistry. 2016;**75**:75-85

[9] Sánchez-López E, Marina ML, Crego AL. Improving the sensitivity in chiral capillary electrophoresis. Electrophoresis. 2016;**37**:19-34

[10] Iadarola P, Fumagalli M, Bardoni AM, Salvini R, Viglio S. Recent applications of CE and HPLC-MS in the analysis of human fluids. Electrophoresis. 2016;**37**:212-230

[11] Scriba GKE. Recent advances in peptide and peptidomimetic stereoisomer separations by capillary electromigration techniques. Electrophoresis. 2006;**27**:222-230

[12] Szumski M, Buszewski B. State of the art in miniaturized separation techniques. Critical Reviews in Analytical Chemistry. 2002;**32**:1-46

[13] Rieux L, Sneekes E-J, Swart R. Nano LC: Principles, evolution, and state-of-the art of the technique. LC-GC North America. 2011;**29**:926-934

[14] Zotou A. An overview of recent advances in HPLC instrumentation. Central European Journal of Chemistry. 2012;**10**:554-569

[15] Fanali C, Dugo P, Dugo L, Mondello L. Capillary-liquid chromatography (CLC) and nano-LC in food analysis. Trends in Analytical Chemistry. 2013:3-4

[16] Ahmed M, Yajadda MM, Han ZJ, Su D, Wang G, Ostrikov KK, et al. Single-walled carbon nanotube-based polymer monoliths for the enantioselective nano-liquid chromatographic separation of racemic pharmaceuticals. Journal of Chromatography. A. 2014;**1360**:100-109

[17] Ciogli A, Pierri G, Kotoni D, Cavazzini A, Botta L, Villani C, et al. Toward enantioselective nano

ultrahigh-performance liquid chromatography with whelk-O1 chiral stationary phase. Electrophoresis. 2014;**35**:2819-2823

[18] Fanali S, D'Orazio G, Farkas T, Chankvetadze B. Comparative performance of capillary columns made with totally porous and core-shell particles coated with a polysaccharidebased chiral selector in nano-liquid chromatography and capillary electrochromatography. Journal of Chromatography. A. 2012;**1269**:136-142

[19] Aturki Z, D'Orazio G, Rocco A, Si-Ahmed K, Fanali S. Investigation of polar stationary phases for the separation of sympathomimetic drugs with nano-liquid chromatography in hydrophilic interaction liquid chromatography mode. Analytica Chimica Acta. 2011;**685**:103-110

[20] Auditore R, Santagati NA, Aturki Z, Fanali S. Enantiomeric separation of amlodipine and its two chiral impurities by nanoliquid chromatography and capillary electrochromatography using a chiral stationary phase based on cellulose tris(4-chloro-3-methylphenylcarbamate). Electrophoresis. 2013;**34**:2593-2600

[21] D'Orazio G, Fanali S. Enantiomeric separation by using nano-liquid chromatography with on-column focusing. Journal of Separation Science. 2008;**31**:2567-2571

[22] Heus F, Giera M, de Kloe GE, van Iperen D, Buijs J, Nahar TT, et al. Development of a microfluidic confocal fluorescence detection system for the hyphenation of nano-LC to on line biochemical assays. Analytical and Bioanalytical Chemistry. 2010;**398**:3023-3032

[23] D'Orazio G, Rocco A, Fanali S. Fast-liquid chromatography using columns of different internal diameters packed with sub-2μm silica particles. Journal of Chromatography. A. 2012;**1228**:213-220

[24] Vega ED, Crego AL, Lomsadze K, Chankvetadze B, Marina ML. Enantiomeric separation of FMOC-amino acids by nano-LC and CEC using a new chiral stationary phase, cellulose tris(3-chloro-4-methylphenylcarbamate). Electrophoresis. 2011;**32**:2700-2707

[25] Zheng M, Wu Y, Lu L, Ding K, Tang F, Lin Z, et al. Simultaneous analysis of acetaminophen, p-aminophenol and aspirin metabolites by hydrophilic interaction and strong anion exchange capillary liquid chromatography coupled to amperometric detection. Journal of Separation Science. 2011;**34**:2072-2078

[26] Pröfrock D. Progress and possible applications of miniaturized separation techniques and elemental mass spectrometry for quantitative, heteroatom-tagged proteomics. Analytical and Bioanalytical Chemistry. 2010;**398**:2383-2401

[27] She YF, Smith RD. Advanced nanoscale separations and mass spectrometry for sensitive high-throughput proteomics. Expert Review of Proteomics. 2005;**2**:431-447

[28] Mateos J, Lourido L, Fernández-Puente P, Calamia V, Fernández-López C, Oreiro N, et al. Differential protein profiling of synovial fluid from rheumatoid arthritis and osteoarthritis patients using LC–MALDI TOF/TOF. Journal of Proteomics. 2012;**75**:2869-2878

[29] Joseph R, Srivastava OP, Pfister RR. Differential epithelial and stromal protein profiles in keratoconus and normal human corneas. Experimental Eye Research. 2011;**92**:282-298

[30] Duan X, Dai L, Chen S-C, Balthasar JP, Qu J. Nano-scale liquid chromatography/mass spectrometry and on-the-fly orthogonal array optimization for quantification of therapeutic monoclonal antibodies and the application in preclinical analysis. Journal of Chromatography A. 2012;**1251**:63-73

[31] Hernáez ML, Ximénez-Embún P, Martínez-Gomariz M, Gutiérrez-Blázquez MD, Nombela C, Gil C. Identification of *Candida albicans* exposed surface proteins in vivo by a rapid proteomic approach. Journal of Proteomics. 2010;**73**:1404-1409

[32] Marie B, Zanella-Cléon I, Guichard N, Becchi M, Marin F. Novel proteins from the calcifying shell matrix of the pacific oyster *Crassostrea gigas*. Marine Biotechnology. 2011;**13**:1159-1168

[33] Javier S, Barbara R, Javier B, Maria A, Miguel D. Evaluation of two molecularly imprinted polymers for the solid-phase extraction of natural, synthetic, and mycoestrogens from environmental water samples before liquid chromatography with mass spectrometry. Journal of Separation Science. 2015;**35**:2692-2699

[34] Rahayu A, Wah Lim L, Takeuchi T. Polymer monolithic methacrylate base modified with tosylated-polyethylene glycol monomethyl ether as a stationary phase for capillary liquid chromatography. Talanta. 2015;**134**:232-238

[35] Upadhyay R, Roy S, Dalwadi M. Umesh Upadhyay a review of advance study on nano liquid chromatography and its application: The future of the field. International Journal of Creative Research Thoughts. 2022;**10**:392-395

Chapter 4

Two-Dimensional Liquid Chromatography Advancing Metabolomics Research

Yatendra Singh and Sixue Chen

Abstract

Multidimensional separation systems offer several advantages over traditional one-dimensional separation systems, particularly their ability to separate molecules from complex mixtures. Two-dimensional liquid chromatography (2D-LC) significantly enhances the ability to analyze complex mixtures by providing greater separation power, sensitivity, and flexibility, making it an invaluable tool for metabolomics research. The 2D-LC is an exciting mode when pursuing untargeted analysis, as it allows for high-resolution separation and subsequent identification and quantification of more analytes. This chapter summarizes the current applications of 2D-LC in metabolomics and the setups of different separation modes that are being employed, presenting the most suitable combinations of chromatographic methods for different targeted and untargeted metabolomics applications.

Keywords: liquid chromatography, two-dimensional liquid chromatography, mass spectrometry, metabolomics, artificial intelligence

1. Introduction

Metabolomics is the comprehensive study of metabolites within biological systems. Metabolites are small molecules that are intermediates and products of metabolism. The metabolome is closest to the phenotype of a biological system (cells, tissues, organs, or biological fluids) and provides a snapshot of the metabolic state of a system at a specific time. Metabolomics involves identifying and quantifying metabolites to understand metabolic processes and their alterations under various conditions [1]. Comprehensive analysis of the metabolome is challenging due to its diverse physicochemical properties, from very hydrophilic to lipophilic and neutral to multiply charged, necessitating a combination of different analytical techniques to provide comprehensive coverage of the metabolites.

Analytical techniques like gas chromatography (GC), capillary electrophoresis (CE), and liquid chromatography (LC) are commonly used to separate metabolites. Among them, LC is superior in analyzing a wide range of metabolites. However, LC cannot easily separate many metabolites from complex mixtures due to co-elution and often cannot separate isomers like Leucine and isoleucine. Sometimes peaks may overlap with other relevant analytes or matrix compounds. When dealing with complex mixtures and the need for high-resolution separation, two-dimensional liquid chromatography (2D-LC) is an attractive approach. The 2D-LC coupled with mass spectrometry (MS) allows for high-throughput and high-resolution analysis of metabolites [2]. It ensures that metabolites enter MS with minimal interference, improving the accuracy of MS analysis. Recently, 2D-LC has gained popularity in metabolomics research and metabolomics has a tremendous application that covers a wide range of samples (e.g. plasma, cosmetics, fecal, urinary, and plant extracts) [3]. 2D-LC becomes especially powerful when handling highly complex samples. For this, different 2D-LC combinations of separation modes like HILIC × HILIC, HILIC × RPLC, RPLC × RPLC, NPLC × RPLC, RPLC × SFC, Achiral LC × Chiral LC, HILIC × SFC, and so forth (HILIC, hydrophilic interaction chromatography; RPLC, reversed-phase liquid chromatography; NPLC, normal-phase liquid chromatography; SFC, supercritical fluid chromatography) were introduced and hyphenated with MS. In 2D-LC, both separations are orthogonal, meaning that separation mechanisms in the first dimension (^{1}D) and second dimension (^{2}D) are significantly different. Currently, 2D-LC has predominantly been used in clinical research, pharmaceutical [4, 5], food [6, 7], and plant biology [8], toxicology [9], and environmental research [10] are widespread applications and are gaining popularity [11, 12].

Current trends in chromatographic prediction using artificial intelligence (AI) and machine learning (ML) algorithms enable fast and accurate predictions of chromatographic behavior [13]. AI-driven retention time (RT) prediction models can accurately estimate RTs, by analyzing molecular properties and experimental conditions, thereby aiding metabolomics research. The AI deep learning techniques like convolutional neural networks (CNNs) and recurrent neural networks (RNNs) can analyze chromatograms and identify peaks, patterns, and anomalies with high precision. They facilitate automated peak integration, deconvolution, and noise reduction, leading to improved quantification accuracy [14].

This chapter focuses on the recent application of 2D-LC in metabolomics research. LC × LC is the hyphenation of two chromatography systems, and it is used for isomer separation, screening, and profiling because it enables the separation of different classes of metabolites and their constituents. Moreover, the integration of AI and 2D-LC MS/MS hyphenation that permits fragmentation is a useful tool for achieving global metabolic profiling of various types of biological samples.

2. Principle of 2D-LC and its characteristics

The 2D-LC can be broadly categorized into offline and online configurations. Offline 2D-LC involves completing ^{1}D, collecting the fractions manually, followed by injecting these fractions into ^{2}D for further separation. This configuration appears to be laborious and time-consuming, involves manual handling of the collected fractions, and carries a risk of sample loss. On the other hand, this configuration provides greater flexibility in method development, independent optimization, and separation efficiency. Offline setups also have the advantage of preconcentration, derivatization, and reconstitution into an appropriate mobile phase between the two separations [1]. Online 2D-LC involves the direct transfer of eluents through a high-pressure switching valve from the ^{1}D dimension to the ^{2}D without manual intervention.

Online 2D-LC is preferred for high-throughput applications, automation and faster analysis, reduced sample handling, and decreased risk of contamination and degradation [15]. It has been successfully applied to separating and analyzing trace analytes and samples with complex matrices [16, 17]. Online setup requires complex instrumentation, simultaneous method optimization is challenging, and less flexibility in the independent adjustment of conditions and eluate amount.

The method development in 2D-LC is not straightforward. While selecting the ^{1}D and ^{2}D columns, multiple factors should be considered. The chromatographic parameters such as suitable column dimensions (length and diameters), particle sizes, flow rates, and second-dimension injection volumes (i.e., loop sizes) should be optimized in comprehensive 2D-LC [18]. Furusho et al. suggested that a shallow gradient should be used in the ^{1}D [16]. The separation power of the ^{1}D system should be high to separate as many metabolites as possible, while in the ^{2}D, separation should be efficient, quick, and compatible with the detector attached to the system.

The distribution of various separation mechanisms employed in the ^{1}D and ^{2}D was discussed by Wei-Bin et al. [9]. RPLC remains the most popular mechanism, and it is utilized in 46 and 75% of ^{1}D and ^{2}D separations, respectively. Of all online 2D separations, approximately 33% use RPLC in both dimensions and about 33% of all online 2D [11]. Various types of mass spectrometry (72.1%) are the dominant detection techniques [8], for which volatile solvents (e.g., acetonitrile and methanol) and volatile salts (e.g., ammonium acetate) are often preferred for LC.

2D-LC allows the detection of low-abundance metabolites in complex mixtures that may be missed in 1D-LC. It can differentiate isomers by separating them based on different properties in each dimension, which is often challenging for 1D-LC. Moreover, it reduces matrix effects and interferences from other metabolites in the sample that might affect the metabolite identification and quantification. 2D-LC combines different separation systems like reverse phase chromatography, normal phase chromatography, volume exclusion chromatography, affinity chromatography, and ion exchange chromatography to resolve the chromatographic separation. It also contributes to the enrichment of target metabolites, and abundant sample

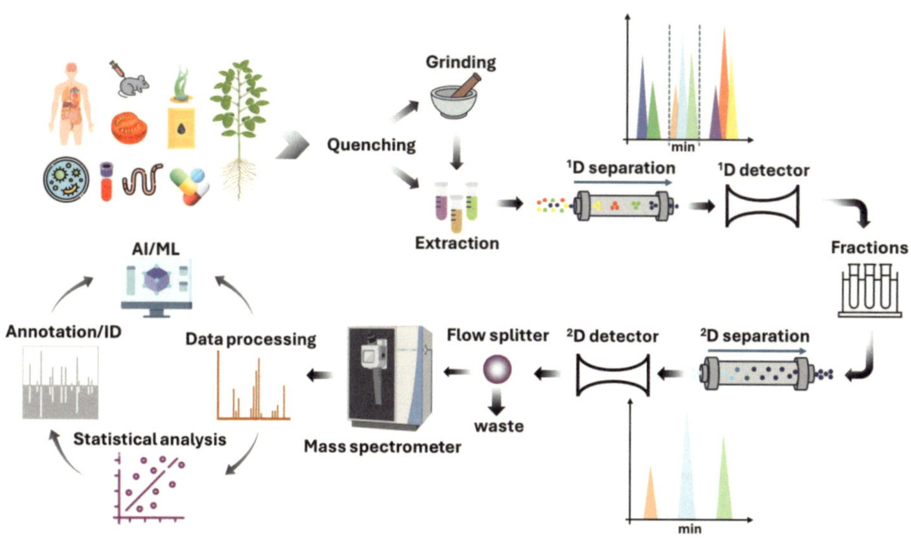

Figure 1.
Major 2D-LC applications utilized in metabolomics research.

information can be obtained efficiently. It has become a popular technique in various fields (**Figure 1**) owing to its unique advantages, but it also increases the cost of the instrument to a certain extent.

3. Detection sensitivity of 2D-LC methods

Although excellent separation of complex mixtures was achieved by 2D-LC, detection sensitivity is another aspect of performance that requires serious attention in metabolomics research. The dilution problem arises in 2D-LC because analytes eluted from the ^{1}D column may be diluted during the ^{2}D separation before reaching the detector. This resulted in a lower concentration of each analyte reaching the detector. It directly affects the ability to detect and quantify trace amounts of metabolites. Insufficient sensitivity has long been seen as a major limitation of 2D methods, and this concern is justified in applications with limited sample amounts. To address the sensitivity issue, researchers often employ pre-concentration techniques like solid-phase extraction (SPE), or liquid-liquid extraction (LLE), and specific detection techniques like MS and modulation techniques. Temperature-responsive liquid chromatography (TRLC) in the ^{1}D hyphenated with RPLC in ^{2}D solves the dilution problem and enhances sensitivity *via* exploiting the solute refocusing effect [19, 20]. The solvent mismatch problem is another challenge associated with 2D-LC [5]. In recent years, various active modulation approaches have been developed to address the mismatch problem between ^{1}D effluent and the ^{2}D mobile phase. Since the introduction of stationary phase-assisted modulation (SPAM), it has become a widely used method [21]. Later, active solvent modulation (ASM), which is a valve-based approach, was introduced [22]. Schmitz et al. presented the concept of utilizing an auxiliary pump, known as a "transfer pump," to force fractions out of sample loops to mix them with a 2D mobile phase that serves as a diluent [23]. It is much like the previously introduced at-column dilution (ACD), which requires an auxiliary pump, which allows continuous dilution factor variation [24]. To lessen the effects of ^{1}D and ^{2}D mobile phase mismatch, Tang et al. reported the utility of installing an inline mixer between the interface valve and the ^{2}D column [25]. Vacuum-evaporation modulation [26] and thermal modulation [27] are other active modulation techniques. The reader can refer to cited articles for basic principles of various modulation techniques [2, 28–30].

4. Technical advances in 2D-LC

The World Health Organization estimated that approximately 0.4 million people die yearly from ingestion of contaminated food. Analyzing all products where chromatography is applicable is the solution, not just for food safety prevention but also for introducing high-quality products into the supply chain. Current trends of AI and ML in chromatographic prediction can lead to faster and more accurate chromatographic analysis. Recently, Phytocontrol (a French company) has collaborated with Fujitsu (a Japanese company) to automate chromatographic techniques for AI-guided food contaminant analysis [31]. Den Uijl et al. have recently conducted a thorough investigation on retention modeling in conjunction with scanning gradients for LC optimization [32]. This work investigated the feasibility of using the RT data obtained from fast ^{2}D separations to build retention models to guide the development of ^{1}D elution conditions using the same stationary phase. The authors discovered that doing so produced erroneous predictions, most likely because of extrapolation to gradient slopes beyond the range of those involved in the initial data collection using 2D-LC conditions. Subsequently, it was demonstrated by Den Uijl et al. that isocratic retention of analytes can be predicted by scanning-gradient-based models on trap columns under the dilution-flow conditions used in SPAM [32]. The authors emphasized the careful selection of the dilution flow under specific conditions, and dilution with a weaker eluent was shown to be counterproductive [32]. In another study, Boelrijk et al. explained how to apply Bayesian optimization for 2D-LC optimization of various parameters derived from experiments [33]. Molenaar et al. developed a peak-tracking algorithm for LC × LC and applied it to peptide retention data obtained under different chromatographic conditions [34]. Retention modeling was used to expedite the development of 2D-LC methods for chiral and achiral pharmaceutical separations [35, 36]. In these studies, the retention data obtained from 1D-LC or 2D-LC were used to train 2D retention models. The resulting models were then used to develop resolution maps depending on various variables, such as column temperature and mobile phase composition. The generated resolution maps can be used to determine the circumstances that are most likely to produce key peak pairs in the ^{2}D with appropriate resolution. Stoll and Pirok presented retention modeling for 2D-LC and concluded that it can decrease the amount of trial and error required [37]. Subsequently, they developed a 2D-LC-MS method using this model to discover impurities co-eluting with primary peaks in 1D-LC therapeutic peptide analysis [38, 39].

Artificial neural network (ANN) was reported to be the most effective and accurate for predicting LC chromatographic characteristics. The ANN combined with the quantitative structure retention relationship (QSRR) was used for RT prediction [13]. Therefore, AI algorithms are powerful tools that automate tasks and enhance the precision of analytical methods, especially in metabolomics. The 2D-LC is an effective tool for separating and identifying metabolites, but it generates massive and complicated datasets, which can be challenging. Therefore, AI algorithms are usually used to analyze these datasets and identify patterns and trends that would be difficult to find through manual inspection. Signal preprocessing is primarily concerned with individual chromatographic signals. It addresses problems such as modulation-phase adjustment [40, 41], peak identification [42], noise removal [43], baseline drift correction [44], and peak distortions [45]. Its main goal is to improve the raw 2D-LC data analysis. Additionally, in data processing, the ROIMCR (regions of interest multivariate curve resolution) approach [46] is a sensitive and adaptable technique for identifying and quantifying a broad spectrum of metabolites in complex samples [47]. Recently, software "Program for Interpretive Optimization of Two-dimensional Resolution" (PIOTR) developed by Pirok et al. [48] has shown to speed up LC × LC method development, based on only a few experiments with the consideration of retention behavior of the analytes under varying isocratic or gradient mobile-phase conditions. Therefore, it is anticipated that these algorithms are expected to be applied to ensure the accuracy of individual peaks. Several AI technologies have been developed to optimize signal preprocessing strategies, improve the quality of individual signals before data integration, and handle the complexities of each chromatographic trace [48, 49]. Furthermore, large amounts of data produced by metabolomics studies were analyzed using ML-based techniques to identify patterns and distinguish constituents within their classes by assessing features specific to each class [23]. These techniques include cluster analysis, principal component analysis, hierarchical clustering, uniform manifold approximation and projection, t-distributed stochastic neighbor embedding, k-nearest neighbors, fisher ratio, naïve Bayes, logistic regression, linear discriminant analysis, partial least squares, partial least squares-discriminant analysis, and quadratic discriminant analysis [23–25].

The integration of AI into the development and optimization of 2D-LC represents a significant technological advancement, offering enhanced capabilities, efficiency, and accuracy in the analysis of complex mixtures. Despite these advances, challenges remain. Data quality and quantity are critical for successful AI and ML applications. Hybrid models, integrating one or more AI or ML models, can be used to predict different chromatographic characteristics. An ongoing trend is the creation of large, well-curated chromatographic databases that facilitate model learning, training, and validation. As AI continues to evolve, its applications in 2D-LC are expected to expand, providing even greater benefits to analytical chemistry.

5. Experimental design, sample preparation, and separation technologies

An effective metabolite extraction procedure is the most critical step in obtaining high-quality metabolomics data. To achieve this, sample quenching was performed in liquid nitrogen, followed by homogenization (**Figure 2**). After homogenization, several selective metabolite extraction methods like microwave-assisted extraction, ultrasound-assisted extraction, high-voltage electric discharge extraction, supercritical fluid extraction, enzyme-assisted extraction, and SPE were utilized, which are extensively covered in cited references [50–52]. Combining two or more separation techniques in a single analysis can also reduce the co-elution of metabolites and improve separation efficiency. Some examples include 2D gas chromatography (2D-GC), supercritical fluid chromatography (2D-SFC), and two- and three-dimensional LC (2D-LC and 3D-LC) [53]. Among them, 2D-LC exhibits superior chromatographic

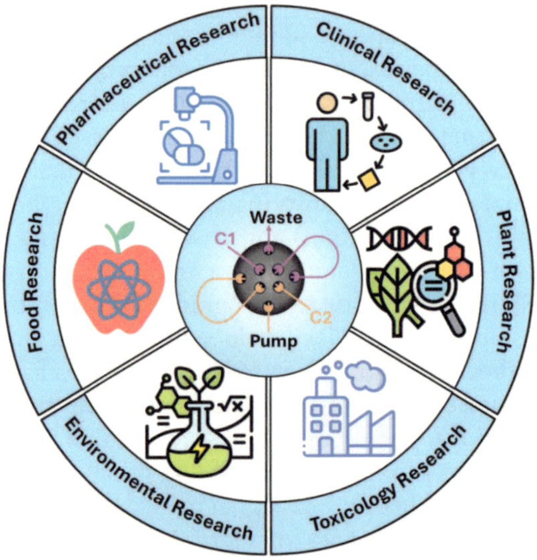

Figure 2.
A general workflow for 2D-LC-based metabolomics research, including sample preparation, two-dimensional separation, MS data acquisition, and analysis. Please note that for online 2D-LC, the [1]D detector and fractions are eliminated.

capabilities, including enhanced resolution, peak capacity, and sensitivity, and covers a wide range of metabolites. It has been extensively utilized for untargeted metabolomics. Different 2D-LC combinations of chromatography like HILIC × HILIC, HILIC × RPLC, RPLC × RPLC, NPLC × RPLC, RPLC × SFC, Achiral LC × Chiral LC, HILIC × SFC, and so on were introduced and hyphenated with various types of MS detectors. Lv et al. used parallel column 2D-LC-MS for broad coverage of metabolome and lipidome [54]. Online 2D-LC offers high-throughput and automated analysis, although ^{2}D chromatographic resolution can be reduced by short separation times.

6. Application of 2D-LC in metabolomics research

2D-LC is a powerful analytical technique increasingly used in metabolomics research due to its enhanced separation capabilities, sensitivity, and resolution. 2D-LC allows detailed profiling of polar to nonpolar metabolites (**Figure 1**). This section summarizes recent applications of 2D-LC in **Table 1** and discusses them

Sample type	1D	2D	Detection technique	Analytical purpose of the application and its references
Rice's shoots	HILIC	RP	micrOToF-Q II	Untargeted metabolomics of rice shoots [55]
Plasma, seminal urine, and fecal	HILIC	RP	TQD-MS	Targeted metabolomic analysis [56]
Seeds	HILIC	RP	QToF-MS	Untargeted metabolomics of *Cuscuta chinensis* [57]
Human plasma	HILIC	RP	QToF-MS	Lipid analysis [58]
Cosmetics	HILIC	RP	TqD-MS	Detection of prohibited substances in cosmetics [59]
Drug mixtures	HILIC	RP	TQD-MS	Quantification of amino acids in drug mixtures [60]
Caenorhabditis elegans	HILIC	RP	Orbitrap-MS	Identification and quantification of Cardiolipin [61]
Xanthoceras sorbifolium	HILIC	RP	Orbitrap-MS	Targeted discovery of novel barrigenol-type triterpenoid saponins [62]
Fecal samples	HILIC	RP	QToF-MS	Untargeted metabolite characterization [63]
Plasma	HILIC	RP	Q-ToF-MS	Metabolites and lipid analysis [64]
Human plasma	HILIC	RP	ESI-MS	Quantification of steroid hormones [65]
Herbal samples	HILIC	RP	QToF-MS	Triterpene and saponins characterization [66]
Plasma samples	HILIC	RP	QToF-MS	Metabolite analysis in human plasma [67]
Mice blood	HILIC	RP	QTrap-MS	Screening of polar and nonpolar metabolites [68]
Fermentation broth	RP	RP	PDA-ESI-MS	Microspheres of shikimate-producing *Escherichia coli* [69]
Root	RP	RP	ESI-MS	Untargeted metabolomic analysis [70]
Soy sauce	RP	RP	QToF-MS	Metabolite identification and structure analysis [71]

Sample type	1D	2D	Detection technique	Analytical purpose of the application and its references
Plant sample	RP	RP	Orbitrap-MS	Molecular networking of metabolites in *Yupingfeng* [72]
Fabric softener	RP	RP	cIM-MS	Identification of environmental toxicants [73]
Human urine	RP	RP	LTQ-Orbitrap-MS	Comprehensively profiling and functionally annotation of the metabolome [74]
Plant sample	RP	RP	IM-QToF-MS	Metabolite characterization from three *Glycyrrhiza* species [75]
Fatty acids	RP	RP	DAD-QToF-MS	Profiling of conjugated fatty acid isomers [76]
Roots	RP	RP	ESI-QToF-MS	Screening of bioactive compounds [77]
Leaves	RP	RP	TQD-MS	Analysis of 17-Hydroxygeranyllinalool diterpene glycosides in *Nicotiana tabacum* [78]
Leaves	RP	RP	PDA-TQD-MS	Untargeted metabolite characterization [53]
Roots	RP	RP	QToF-MS	In-depth metabolite characterization [53]
Serum samples	RP	NP	Q-ToF-MS	Screening of pesticides, veterinary/human drugs, and other chemical pollutants [79]
Human plasma	NP	RP	QToF-MS	Biomarkers for disease diagnosis [80]
Leaves & roots	NP	RP	Orbitrap-MS	Structural analysis of prenylated phenolic compounds [81]
Amino acids	Achiral	Chiral	QToF-MS	Untargeted enantioselective amino acid analysis [82]
HCA-7 cells	Achiral	Chiral	QqQ-MS	Enantioselective quantitation of oxylipins [83]
Amino acids	Achiral	Chiral	QToF-MS	Enantioselective separation of amino acids [84]
Herbs	SEC	HILIC	CAD	Separation and analysis of polysaccharides [85]
Aqueous film-forming foams	RPWAX	RP	Orbitrap-MS	Untargeted identification of poly- and perfluoroalkyl substances [86]
Sugar phosphates	HILIC	HILIC	QToF-MS	Analysis of sugar phosphates in glycolysis and pentose phosphate pathways [87]

Abbreviations: reversed-phase weak anion exchange (RPWAX), reverse phase (RP), hydrophilic interaction chromatography (HILIC), normal phase (NP), supercritical fluid chromatography (SFC), triple quadrupole detector (TQD), electrospray ionization (ESI), mass spectrometry (MS), photodiode array (PDA), cyclic ion-mobility (cIM), diode array detector (DAD), quadrupole time-of-flight (QToF), charged aerosol detector (CAD), and size-exclusion chromatography (SEC).

Table 1.
A brief overview of recent 2D-LC applications in metabolomics research.

in the text. The following parameters, such as sample type, ^{1}D and ^{2}D, detection technique, and analytical purpose are included in each application (**Table 1**). The recent applications of 2D-LC are explored in clinical research, pharmaceutical research, food research, plant sciences, toxicology research, and environmental research (**Figure 1**).

HILIC offers significant advantages when combined with RPLC for metabolomics by providing a valuable strategy to enhance metabolite coverage and separation efficiency. RPLC is widely used due to its reproducibility and reliability, whereas HILIC enhances metabolite coverage, especially for highly polar compounds. Combining HILIC × RPLC allows for simultaneous analysis of hydrophilic and hydrophobic metabolites in a single injection [88–90], yielding a broader metabolome coverage due to the complementary separation mechanisms. Additionally, HILIC × RPLC enhances separation efficiency, peak capacity, and overall analytical performance in untargeted metabolomics, providing a deep insight into the metabolic changes associated with health and disease states [91]. The HILIC × RPLC was tested using several biological samples, that is, plasma, serum, urine, fecal, seminal plasma, and liver. Guo et al. used this system to measure 417 metabolites in plasma, serum, urine, fecal, and seminal plasma, covering the polar to the nonpolar range [56]. Wu et al. reported that 331 metabolite features were identified using the offline HILIC × RPLC from *Lilium lancifolium* and *L. brownii*, which are commonly used in the production of Chinese medicine and vegetables [92]. Using HILIC × RPLC coupled with HR-MS, 302 metabolites were structurally identified or tentatively characterized from *Cuscuta chinensis* [57]. Zhang et al. devised a HILIC × RPLC approach based on dilution modulation that allows for the simultaneous detection of 126 prohibited substances in cosmetic products [59]. The HILIC × RPLC has also been used for targeted analysis of specific metabolites. For example, Helmer et al. recently utilized the HILIC × RPLC method to separate cardiolipins and their oxidation products [93]. Dang et al. conducted a target separation of flavonoids from *Saxifraga tangutica* [94]. The technology is also applied in the simultaneous analysis of highly polar and nonpolar pesticides in food [55]. Laan et al. developed a comprehensive identification workflow to identify taste-related RT and m/z features of unknown compounds in soy sauce [71].

The hyphenation of HILIC × RPLC with HR-MS is advantageous for lipidomics. Separation according to the head group can be achieved using HILIC, which allows the collection of individual classes. Then, the lipids can be further separated according to fatty acid chain properties using RPLC. Sorensen et al. implemented HILIC in the ^{1}D to fractionate lipids in human plasma into nine fractions and subjected them to a ^{2}D for further separation using a C18 column to resolve over 1000 lipids from human plasma [58]. To tackle the challenge of multiple lipids co-eluting from LC causing ion suppression in MS, Xu et al. developed Lipid Wizard software to analyze lipid profiles acquired by HILIC × RPLC-MS, leading to high confidence in lipid assignment and high accuracy of lipid quantification [95].

Recently, a synthesized monolithic column composed of a copolymer of styrene, divinyl benzene, and 1-vinyl-1,2,4-tria-zole has allowed for the separation of both polar and nonpolar metabolites [68]. A new mixed-mode stationary phase derived from [2-(3, 4-epoxycyclohexyl) ethyl] trimethoxysilane offers hydrophilic interaction, reversed-phase, and ion-exchange functionalities, which facilitate the simultaneous separation of polar, nonpolar, and ionic compounds [91]. These advancements have great potential in pushing biomolecular separation to the next level of frontiers.

Another popular 2D-LC method is RPLC × RPLC, which offers wide applicability of reverse-phase conditions to separate metabolites with high efficiency and great resolution. In addition to manipulating the functional group chemistry of the stational phase, high pH and low pH fractionation are often utilized to change the chemical properties of analytes and achieve high-resolution separation. With HILIC fractionation of metabolite extracts from three *Astragalus* species (*A. membranaceus* var. *mongholicus*, and *A. membranaceus*), an online RPLC × RPLC was used to separate and characterize 513 metabolites from the fractionated samples [53]. The RPLC × RPLC technique increases peak capacity and enables the resolution of potential co-eluting metabolites, leading to the detection of a higher number of metabolites than the conventional 1D-LC system. For example, Wong et al. report that 120 metabolites were detected using the RPLC × RPLC system from *Glycyrrhiza glabra* extract, and compared with 1D-LC, 2-time more metabolites were separated using 2D-LC [70]. Xu et al. used 1D-LC and RPLC × RPLC LC to identify 2357 metabolites in normal human urine. The 1D-LC-MS/MS technique identified lipid and amino acids, whereas the 2D-LC-MS/MS technique profiled additional metabolites to lipid and amino acids [74]. Liu et al. used ion exchange chromatography in ^{1}D separation to fractionate between acidic and weakly acidic components. These subfractions were then separated by ^{2}D and ^{3}D chromatography using either online RPLC × RPLC or offline HILIC × RPLC. As a result, 1097 metabolites were identified from three *Glycyrrhiza* species [75].

NPLC × RPLC is another popular 2D-LC combination that offers high resolution for polar metabolites in ^{1}D and nonpolar metabolites in ^{2}D. NPLC separates compounds in a mixture based on their polarity using columns packed with polar stationary phases like silica gel or alumina and nonpolar or moderately polar mobile phases. It can be used for a wide range of applications, from small organic molecules to larger natural products (**Figure 1**). This technique is also useful for neutral metabolites, differing in hydrophobicity and polarity. Yang et al. demonstrated the promising potential of NPLC × RPLC 2D-LC-QToF-MS for identifying 13 lipid species that were proposed as potential lipid biomarkers for Lacunar infarction [80]. In addition, NPLC × RPLC 2D-LC-Orbitrap-MS coupled with mass defect filter (MDF) technology was used to characterize 1631 prenylated phenolics in a targeted manner [81].

The achiral derivatization of analytes can improve their molecular properties by considering chromatographic separation, ionization stability, and MS detection sensitivity. For example, several researchers performed pre-column derivatization of amino acids, to improve separation and detection sensitivity [96–98]. Generally, direct chiral metabolomics involves chiral stationary phases or chiral derivatizing reagents with polysaccharide derivatives, macrocyclic antibiotics, chiral crown ethers, chiral ion exchangers, and donor-acceptor phases as chiral selectors. For untargeted chiral metabolomics, a unique strategy was developed based on the simultaneous chiral derivatization of hydroxy/amine moiety-containing metabolites, including all hydroxy acids and amino acids, with the enantiomeric pair of diacetyl-tartaric anhydride. This method separated 214 chiral compounds, including 106 amino acids and 28 hydroxy acids [99]. To achieve better metabolite coverage, a 3D-LC system equipped with reversed-phase, anion-exchange, and enantioselective columns has been developed to measure trace levels of D-asparagine, D-serine, D-alanine, and D-proline, potential biomarkers of chronic kidney disease in human plasma [16]. Due to the structural similarities of cannabinoids, their purification remains a bottleneck. Offline 2D-semipreparative chromatography (macroporous resin column in ^{1}D and C18 column in ^{2}D) employed sequential processes for the scalable purification of cannabinoids from ethanolic extracts of cannabis inflorescence [100].

The LC × SFC or SFC × LC integration exhibits interesting orthogonality for separating ionizable and neutral molecules. Sarrut et al. demonstrated that RPLC × SFC can generate a slightly higher peak capacity than RPLC × RPLC [101]. It has been proven a desirable tool for low-to-moderate molecular weight and thermally labile metabolites, which cannot be separated by GC and CE. Specifically, 2D SFC employing supercritical CO_2 as the primary mobile phase has polarity like hexane that can be adjusted using polar organic solvents (modifier) as the mobile phase. SFC is a green technology because of its low consumption of organic solvents [102]. The unique properties of supercritical CO_2 include its high density, low viscosity, good diffusivity, and outstanding solvating power, enabling quick and high-resolution investigation using SFC [103]. It has been applied to several areas like pharmaceuticals, pesticides, foods, herbicides, and fossil fuels [104].

Lignin depolymerization produces numerous low-molecular-weight phenolic compounds *via* selective bond cleavage, which can be further used for valuable chemical production. Tammekivi et al. developed a novel online RPLC × SFC method with a trapping column interface for separating phenolic compounds in depolymerized lignin samples [105]. Moreover, an offline 2D RPLC × SFC coupled with Q-ToF-MS/MS was applied to the untargeted analysis of depolymerized lignin. The monomers, dimers, trimers, and tetramers of lignin were separated based on the number of hydroxyl groups and steric effects; as a result, 471 metabolites were detected [105]. Moreover, the SFC method presents special benefits for separating isomers of natural products due to the high degree of orthogonality and significant peak capacity offered by the RP × SFC. Qu et al. identified 324 sesquiterpene alkaloid isomers using RP × SFC, reflecting their separation ability for efficient isomers characterization [106].

Other applications of SFCs hyphenated with the RPLC-HILIC system include the hidden target screening of environmental water samples. This serial coupling of multidimensional chromatography provides an important benefit of extending the range of separable and detectable compounds from "non-polar" to "polar" and even "very polar." This study validated 274 environmental compounds by comparing their RTs and masses with those of their reference standard compounds [3]. Wastewater from hydrothermal liquefaction contains many compounds belonging to several classes that are hazardous. Teboul et al. used RPLC × SFC to analyze hydrothermal liquefaction wastewater from algae conversion, illustrating the power of 2D separation [107]. Sarrut et al. employed the RPLC × SFC system to separate neutral compounds especially, aromatic compounds from an aqueous extract of bio-oil [101]. This integrated system is also useful for separating fatty acids from fish oils [108]. Moreover, online RPLC x SFC can significantly improve the characterization of bio-oils containing thousands of compounds covering a very wide range of molecular weights and polarities [109].

As a modern SFC, ultra-high performance SFC (UHPSFC) can dramatically improve repeatability, stability, and reliability compared to conventional SFCs. Recently, the RPLC × UHPSFC system has been applied to characterize bufadienolides in *Venenum Bufonis* and examine the lipidome differences among three different species of ginseng [103, 110]. Despite the apparent promise of coupling LC with SFC for neutral chemical analysis, some drawbacks have been noted. Because of the nature of the SFC mobile phase, online SFC × LC implementation requires a sophisticated link between the two dimensions. The opposite combination LC × SFC experiences severe injection effects, particularly when the LC dimension is operated in a reversed phase. Currently, inadequate commercial instrumentation limits its potential in online mode [111].

Both offline and online 2D LC-based methods were used to examine toxicants in household products. Both methods identified the same number of toxicants in the sample. Conventionally, the offline 2D LC-MS method identified ester unsaturated ester quaternary ammonium compounds (QACs) as causative agents of observed toxicity [73]. The online 2D-LC method exploits the combination of a mixed-mode weak anion exchange-reversed phase (RPWAX) with an octadecyl stationary phase

(C18), separating 24 poly- and perfluorinated compounds according to ionic classes and chain length [86].

Conjugated fatty acids, produced by on oxidation of polyunsaturated fatty acids, are present as contaminants in pharmaceutical lipid formulations. For quality control and impurity profiling, ^{1}D LC may not be the best method due to the structural complexity of the resulting multicomponent samples. Olfert et al. applied the 2D-LC chiral × RPLC method hyphenated with online DAD-UV and QTOF-MS/MS for the separation of conjugated polyunsaturated fatty acid isomers and structurally related (saturated, unconjugated, and oxidized) compounds [76]. Additionally, combining orthogonal LC techniques increases the peak capacity compared with 1D LC, enhances the number of identified lipids, and reduces interfering matrix components. Although processing heavy oil into useable products adds value, it necessitates sophisticated technology and thorough characterization required to maximize the production of the most profitable products. Separation of heavy oil has been achieved using 2D-LC and characterized de-asphalted maltenes [112].

Phosphorylated sugar isomer separation is a highly demanding and active field of study, and various unique and sensitive techniques have been employed. For instance, Su et al. evaluated the ability of HILIC to separate metabolites from the pentose phosphate and glycolysis pathways using a mixed-mode HILIC/strong anion exchange (SAX) approach [87]. Co-elute fructose 6-phosphate and glucose 1-phosphate were separated in ^{2}D using a HILICpak VT50-2D column. It allows undisturbed determination of glycolytic phosphorylated carbohydrate metabolites *via* their chromatographic separation from hexose monophosphate metabolites [87].

7. Conclusions

The inherent complexity of samples containing thousands of metabolites has led to the development of 2D-LC. Recent technological advancements have enhanced throughput and detection sensitivity and have solved the problem of solvent mismatches. Synthesized monolithic and mixed mode stational phase columns have started to show utility in metabolomics. Comprehensive 2D-LC-MS/MS has enabled the profiling of a wide range of metabolites and has proven to be a vital tool in many fields of biology and medicine. It appears to be a powerful technique for both targeted and untargeted metabolomics. AI and ML are emerging frontiers in metabolomics that require more learning, training, and validation to predict different chromatographic and MS characteristics.

Acknowledgements

This material is based upon work supported by the National Science Foundation Plant Genome Research Program under Grant No. 2318746 (S.C.).

Abbreviations

^{1}D	first dimension
1D-LC	one-dimensional liquid chromatography
^{2}D	second dimension

2D-GC	two-dimensional gas chromatography
2D-LC	2D liquid chromatography
2D-LC-MS	2D liquid-chromatography mass spectrometry
2D-SFC	2D supercritical fluid chromatography
3D-LC	three-dimensional liquid chromatography
ACD	at-column dilution
AI	artificial intelligence
ANN	artificial neural network
ASM	active solvent modulation
CE	capillary electrophoresis
CNNs	convolutional neural networks
GC	gas chromatography
HILIC	hydrophilic interaction chromatography
LC	liquid chromatography
ML	Machine learning
MS	mass spectrometry
NPLC	normal-phase liquid chromatography
PIOTR	program for interpretive optimization of two-dimensional resolution
QSRR	quantitative structure retention relationship
RNNs	recurrent neural networks
ROIMCR	regions of interest multivariate curve resolution
RPLC	reverse-phase liquid chromatography
SFC	supercritical fluid chromatography
SPAM	stationary-phase-assisted modulation
TRLC	TEMPERATURE-responsive liquid chromatography

Author details

Yatendra Singh* and Sixue Chen
Department of Biology, University of Mississippi, Oxford, MS, USA

*Address all correspondence to: ysing3@olemiss.edu; yatendra.saph@gmail.com

References

[1] Pandohee J et al. Multi-dimensional liquid chromatography and metabolomics, are two dimensions better than one? Current Metabolomics. 2015;**3**(1):10-20

[2] Pirok BW et al. Recent developments in two-dimensional liquid chromatography: Fundamental improvements for practical applications. Analytical Chemistry. 2018;**91**(1):240-263

[3] Bieber S et al. RPLC-HILIC and SFC with mass spectrometry: Polarity-extended organic molecule screening in environmental (water) samples. Analytical Chemistry. 2017;**89**(15):7907-7914

[4] Iguiniz M et al. Two-dimensional liquid chromatography in pharmaceutical analysis. Instrumental aspects, trends and applications. Journal of Pharmaceutical and Biomedical Analysis. 2017;**145**:482-503

[5] Iguiniz M et al. Comprehensive two dimensional liquid chromatography as analytical strategy for pharmaceutical analysis. Journal of Chromatography A. 2018;**1536**:195-204

[6] Liang L et al. Recent development of two-dimensional liquid chromatography in food analysis. Food Analytical Methods. 2022:1-12

[7] Montero L et al. Two-dimensional liquid chromatography approaches for food authenticity. Current Opinion in Food Science. 2023;**51**:101041

[8] Rausch A-K et al. Development, validation, and application of a multi-method for the determination of mycotoxins, plant growth regulators, tropane alkaloids, and pesticides in cereals by two-dimensional liquid chromatography tandem mass spectrometry. Analytical and Bioanalytical Chemistry. 2021;**413**(11):3041-3054

[9] Wei-Bin X et al. Application of two-dimensional liquid chromatography in bioanalysis of drugs and toxicants. Chinese Journal of Analytical Chemistry. 2014;**42**(12):1851-1858

[10] McIlvin MR et al. Online nanoflow two-dimension comprehensive active modulation reversed phase–reversed phase liquid chromatography high-resolution mass spectrometry for metaproteomics of environmental and microbiome samples. Journal of Proteome Research. 2021;**20**(9):4589-4597

[11] Van den Hurk RS et al. Recent trends in two-dimensional liquid chromatography. TrAC Trends in Analytical Chemistry. 2023:117166

[12] Jones O. Two-Dimensional Liquid Chromatography: Principles and Practical Applications. Springer Nature; 2020

[13] Singh YR et al. Current trends in chromatographic prediction using artificial intelligence and machine learning. Analytical Methods. 2023;**15**(23):2785-2797

[14] Caratti A et al. Boosting comprehensive two-dimensional chromatography with artificial intelligence: Application to food-omics. TrAC Trends in Analytical Chemistry. 2024:117669

[15] Stoll DR et al. Two-dimensional liquid chromatography: A state of the

art tutorial. Analytical Chemistry. 2017;**89**(1):519-531

[16] Furusho A et al. Three-dimensional high-performance liquid chromatographic determination of Asn, Ser, Ala, and Pro enantiomers in the plasma of patients with chronic kidney disease. Analytical Chemistry. 2019;**91**(18):11569-11575

[17] Bian X et al. Ultrasensitive quantification of trace amines based on N-phosphorylation labeling chip 2D LC-QQQ/MS. Journal of Pharmaceutical Analysis. 2023;**13**(3):315-322

[18] Schoenmakers PJ et al. A protocol for designing comprehensive two-dimensional liquid chromatography separation systems. Journal of Chromatography A. 2006;**1120**(1-2):282-290

[19] Wicht K et al. Enhanced sensitivity in comprehensive liquid chromatography: Overcoming the dilution problem in LC × LC via temperature-responsive liquid chromatography. Analytical Chemistry. 2022;**94**(48):16728-16737

[20] Wicht K et al. Solving the dilution problem in LC× LC via exploitation of the solute refocusing effect in TRLC× RPLC. In: 17th International Symposium on Hyphenated Techniques in Chromatography and Separation Technology (HTC-17). 2022

[21] Vonk RJ et al. Comprehensive two-dimensional liquid chromatography with stationary-phase-assisted modulation coupled to high-resolution mass spectrometry applied to proteome analysis of *Saccharomyces cerevisiae*. Analytical Chemistry. 2015;**87**(10):5387-5394

[22] Stoll DR et al. Active solvent modulation: A valve-based approach to improve separation compatibility in two-dimensional liquid chromatography. Analytical Chemistry. 2017;**89**(17):9260-9267

[23] Chen Y, Li J, et al. Development of an at-column dilution modulator for flexible and precise control of dilution factors to overcome mobile phase incompatibility in comprehensive two-dimensional liquid chromatography. Analytical Chemistry. 2019;**91**(15):10251-10257

[24] Blom KF et al. Two-pump at-column-dilution configuration for preparative liquid chromatography–mass spectrometry. Journal of Combinatorial Chemistry. 2002;**4**(4):295-301

[25] Tang S et al. Resolving solvent incompatibility in two-dimensional liquid chromatography with In-line mixing modulation. Analytical Chemistry. 2022;**94**(46):16142-16150

[26] Tian H et al. Multidimensional liquid chromatography system with an innovative solvent evaporation interface. Journal of Chromatography A. 2006;**1137**(1):42-48

[27] Niezen LE et al. Thermal modulation to enhance two-dimensional liquid chromatography separations of polymers. Journal of Chromatography A. 2021;**1653**:462429

[28] Stoll DR et al. Multi-Dimensional Liquid Chromatography: Principles, Practice, and Applications. CRC Press; 2022

[29] Chen Y et al. Advance in on-line two-dimensional liquid chromatography modulation technology. TrAC Trends in Analytical Chemistry. 2019;**120**:115647

[30] Chapel S et al. Strategies to circumvent the solvent strength mismatch problem in online

comprehensive two-dimensional liquid chromatography. Journal of Separation Science. 2022;**45**(1):7-26

[31] Bose P et al. Robots in the Lab: AI-Enhanced Chromatography. 2023

[32] den Uijl MJ et al. Assessing the feasibility of stationary-phase-assisted modulation for two-dimensional liquid-chromatography separations. Journal of Chromatography A. 2022;**1679**:463388

[33] Boelrijk J et al. Bayesian optimization of comprehensive two-dimensional liquid chromatography separations. Journal of Chromatography A. 2021;**1659**:462628

[34] Molenaar SRA et al. Peak-tracking algorithm for use in comprehensive two-dimensional liquid chromatography – Application to monoclonal-antibody peptides. Journal of Chromatography A. 2021;**1639**:461922

[35] Haidar Ahmad IA et al. In silico multifactorial modeling for streamlined development and optimization of two-dimensional liquid chromatography. Analytical Chemistry. 2021;**93**(33):11532-11539

[36] Makey DM et al. Mapping the separation landscape in two-dimensional liquid chromatography: Blueprints for efficient analysis and purification of pharmaceuticals enabled by computer-assisted modeling. Analytical Chemistry. 2021;**93**(2):964-972

[37] Stoll D et al. Perspectives on the use of retention modeling to streamline 2D-LC method development: Current state and future prospects. LCGC Supplements. 2022;**40**(s4):30-34

[38] Stoll DR et al. A strategy for assessing peak purity of pharmaceutical peptides in reversed-phase chromatography methods using two-dimensional liquid chromatography coupled to mass spectrometry. Part II: Development of second-dimension gradient conditions. Journal of Chromatography A. 2023;**1693**:463873

[39] Petersson P et al. A strategy for assessing peak purity of pharmaceutical peptides in reversed-phase chromatography methods using two-dimensional liquid chromatography coupled to mass spectrometry. Part I: Selection of columns and mobile phases. Journal of Chromatography A. 2023;**1693**:463874

[40] Milani NB et al. Comprehensive two-dimensional gas chromatography—A discussion on recent innovations. Journal of Separation Science. 2023;**46**(21):2300304

[41] Reichenbach SE et al. GC× GC data visualization, processing, and analysis. In: Comprehensive Analytical Chemistry. Elsevier; 2022. pp. 185-229

[42] Van Stee L et al. Peak detection methods for GC× GC: An overview. TrAC Trends in Analytical Chemistry. 2016;**83**:1-13

[43] Ning X et al. Chromatogram baseline estimation and denoising using sparsity (BEADS). Chemometrics and Intelligent Laboratory Systems. 2014;**139**:156-167

[44] Tian T-F et al. Web server for peak detection, baseline correction, and alignment in two-dimensional gas chromatography mass spectrometry-based metabolomics data. Analytical Chemistry. 2016;**88**(21):10395-10403

[45] Zhang P et al. Application of comprehensive 2D gas chromatography coupled with mass spectrometry in beer and wine VOC analysis. Analytica. 2023;**4**(3):347-373

[46] Gorrochategui E et al. ROIMCR: A powerful analysis strategy for LC-MS metabolomic datasets. BMC Bioinformatics. 2019;**20**(1):256

[47] Pérez-Cova M et al. Untangling comprehensive two-dimensional liquid chromatography data sets using regions of interest and multivariate curve resolution approaches. TrAC Trends in Analytical Chemistry. 2021;**137**:116207

[48] Pirok BW et al. Program for the interpretive optimization of two-dimensional resolution. Journal of Chromatography A. 2016;**1450**:29-37

[49] Fernández-Albert F et al. An R package to analyse LC/MS metabolomic data: MAIT (metabolite automatic identification toolkit). Bioinformatics. 2014;**30**(13):1937-1939

[50] Yan S et al. Recent advances in proteomics and metabolomics in plants. Molecular Horticulture. 2022;**2**(1):17

[51] Vuckovic D. Sample preparation in global metabolomics of biological fluids and tissues. In: Issaq HJ, Veenstra TD, editors. Proteomic and Metabolomic Approaches to Biomarker Discovery. 2nd ed. Boston: Academic Press; 2020. pp. 53-83

[52] Martias C et al. Optimization of sample preparation for metabolomics exploration of urine, feces, blood and saliva in humans using combined NMR and UHPLC-HRMS platforms. Molecules. 2021;**26**(14):4111

[53] Zhao D et al. A multidimensional chromatography/high-resolution mass spectrometry approach for the in-depth metabolites characterization of two astragalus species. Journal of Chromatography A. 2023;**1688**:463718

[54] Lv W et al. Pseudotargeted method based on parallel column two-dimensional liquid chromatography-mass spectrometry for broad coverage of metabolome and lipidome. Analytical Chemistry. 2020;**92**(8):6043-6050

[55] Navarro-Reig M et al. Untargeted comprehensive two-dimensional liquid chromatography coupled with high-resolution mass spectrometry analysis of rice metabolome using multivariate curve resolution. Analytical Chemistry. 2017;**89**(14):7675-7683

[56] Guo R et al. Polarity-extended liquid chromatography-triple quadrupole mass spectrometry for simultaneous hydrophilic and hydrophobic metabolite analysis. Analytica Chimica Acta. 2023;**1277**:341655

[57] Wang M et al. A multi-dimensional liquid chromatography/high-resolution mass spectrometry approach combined with computational data processing for the comprehensive characterization of the multicomponents from cuscuta chinensis. Journal of Chromatography A. 2022;**1675**:463162

[58] Sorensen MJ et al. Two-dimensional liquid chromatography-mass spectrometry for lipidomics using off-line coupling of hydrophilic interaction liquid chromatography with 50 cm long reversed phase capillary columns. Journal of Chromatography A. 2023;**1687**:463707

[59] Zhang L et al. Diluting modulation-based two dimensional-liquid chromatography coupled with mass spectrometry for simultaneously determining multiclass prohibited substances in cosmetics. Journal of Chromatography A. 2023;**1695**:463954

[60] Pérez-Cova M et al. Quantification strategies for two-dimensional liquid chromatography datasets using regions of interest and multivariate

curve resolution approaches. Talanta. 2022;**247**:123586

[61] Helmer PO et al. Investigation of cardiolipin oxidation products as a new endpoint for oxidative stress in *C. elegans* by means of online two-dimensional liquid chromatography and high-resolution mass spectrometry. Free Radical Biology and Medicine. 2021;**162**:216-224

[62] Zhou H et al. A high-efficiency integrated strategy for the targeted discovery of novel barrigenol-type triterpenoid saponins from the shell of *Xanthoceras sorbifolium* Bunge by offline two-dimensional chromatography-Orbitrap mass spectrometry, neutral loss data acquisition, and predicted natural product screening. Arabian Journal of Chemistry. 2024;**17**(1):105445

[63] Anderson BG et al. Offline two-dimensional liquid chromatography–mass spectrometry for deep annotation of the fecal metabolome following fecal microbiota transplantation. Journal of Proteome Research. 2024

[64] Feng J et al. Simultaneous analysis of the metabolome and lipidome using polarity partition two-dimensional liquid chromatography–mass spectrometry. Analytical Chemistry. 2021;**93**(45):15192-15199

[65] Kotasova M et al. A new heart-cutting method for a multiplex quantitative analysis of steroid hormones in plasma using 2D-LC/MS/MS technique. Molecules. 2023;**28**(3):1379

[66] Yasen S et al. Comprehensive characterization of triterpene saponins in rhizoma panacis japonici by offline two-dimensional liquid chromatography coupled to quadrupole time-of-flight mass spectrometry. Molecules. 2024;**29**(6):1295

[67] Orlandi C et al. Miniaturized two-dimensional heart cutting for LC–MS-based metabolomics. Analytical Chemistry. 2023;**95**(5):2822-2831

[68] Basov NV et al. Global LC-MS/MS targeted metabolomics using a combination of HILIC and RP LC separation modes on an organic monolithic column based on 1-vinyl-1, 2, 4-triazole. Talanta. 2024;**267**:125168

[69] Cacciola F et al. Novel comprehensive multidimensional liquid chromatography approach for elucidation of the microbosphere of shikimate-producing *Escherichia coli* SP1. 1/pKD15. 071 strain. Analytical and Bioanalytical Chemistry. 2018;**410**:3473-3482

[70] Wong YF et al. Untargeted profiling of *Glycyrrhiza glabra* extract with comprehensive two-dimensional liquid chromatography-mass spectrometry using multi-segmented shift gradients in the second dimension: Expanding the metabolic coverage. Electrophoresis. 2018;**39**(15):1993-2000

[71] Van der Laan T et al. Fractionation platform for target identification using off-line directed two-dimensional chromatography, mass spectrometry and nuclear magnetic resonance. Analytica Chimica Acta. 2021;**1142**:28-37

[72] Zhu H et al. A compounds annotation strategy using targeted molecular networking for offline two-dimensional liquid chromatography-mass spectrometry analysis: Yupingfeng as a case study. Journal of Chromatography A. 2023;**1702**:464045

[73] Ahmad R et al. Development of a rapid screening method utilizing 2D LC for effect-directed analysis in the identification of environmental toxicants. Science of the Total Environment. 2024;**927**:172199

[74] Xu J et al. A comprehensive 2D-LC/MS/MS profile of the normal human urinary metabolome. Diagnostics. 2022;**12**(9):2184

[75] Liu M et al. Two multidimensional chromatography/high-resolution mass spectrometry approaches enabling the in-depth metabolite characterization simultaneously from three *Glycyrrhiza* species: Method development, comparison, and integration. Journal of Agricultural and Food Chemistry. 2024;**72**(2):1339-1353

[76] Olfert M et al. Comprehensive profiling of conjugated fatty acid isomers and their lipid oxidation products by two-dimensional chiral RP×RP liquid chromatography hyphenated to UV- and SWATH-MS-detection. Analytica Chimica Acta. 2022;**1202**:339667

[77] Guo H et al. Screening and characterization of potential anti-gout components from *Polygonum cuspidatum* by integration off-line two-dimensional liquid chromatography-mass spectrometry with affinity ultrafiltration and on-line HPLC-ABTS. Journal of Pharmaceutical and Biomedical Analysis. 2024;**243**:116103

[78] Chen M et al. Analysis of 17-hydroxygeranyllinalool diterpene glycosides in *Nicotiana tabacum* by using heart-cutting 2D-LC coupled with tandem MS technique. Chromatographia. 2022;**85**(10):931-937

[79] Wang Y et al. Screening strategy for 1210 exogenous chemicals in serum by two-dimensional liquid chromatography-mass spectrometry. Environmental Pollution. 2023;**331**:121914

[80] Yang L et al. Lipidomic analysis of plasma in patients with lacunar infarction using normal-phase/reversed-phase two-dimensional liquid chromatography–quadrupole time-of-flight mass spectrometry. Analytical and Bioanalytical Chemistry. 2017;**409**:3211-3222

[81] Shang Z et al. Characterization of prenylated phenolics in *Glycyrrhiza uralensis* by offline two-dimensional liquid chromatography/mass spectrometry coupled with mass defect filter. Journal of Pharmaceutical and Biomedical Analysis. 2022;**220**:115009

[82] Karongo R et al. Comprehensive online reversed-phase× chiral two-dimensional liquid chromatography-mass spectrometry with data-independent sequential window acquisition of all theoretical fragment-ion spectra-acquisition for untargeted enantioselective amino acid analysis. Analytical Chemistry. 2022;**94**(49):17063-17072

[83] Kampschulte N et al. Deducing Formation Routes of Oxylipins by Quantitative Multiple Heart-Cutting Achiral-Chiral 2D-LC-MS. 2024

[84] Karongo R et al. Enantioselective multiple heart cutting online two-dimensional liquid chromatography-mass spectrometry of all proteinogenic amino acids with second dimension chiral separations in one-minute time scales on a chiral tandem column. Analytica Chimica Acta. 2021;**1180**:338858

[85] Wang H et al. An analysis of polysaccharides from eight plants by a novel heart-cutting two-dimensional liquid chromatography method. Food. 2024;**13**(8):1173

[86] Renai L et al. Development of a comprehensive two-dimensional liquid chromatographic mass spectrometric method for the non-targeted identification of poly- and perfluoroalkyl substances in aqueous film-forming

foams. Analytica Chimica Acta. 2022;**1232**:340485

[87] Su M et al. Isomer selectivity of one- and two-dimensional approaches of mixed-mode and hydrophilic interaction liquid chromatography coupled to tandem mass spectrometry for sugar phosphates of glycolysis and pentose phosphate pathways. Journal of Chromatography A. 2023;**1688**:463727

[88] Hosseinkhani F et al. Systematic evaluation of HILIC stationary phases for global metabolomics of human plasma. Metabolites. 2022;**12**(2):165

[89] Pičmanová M et al. Rapid HILIC-Z ion mobility mass spectrometry (RHIMMS) method for untargeted metabolomics of complex biological samples. Metabolomics. 2022;**18**(3):16

[90] Pirttilä K et al. Automated sequential analysis of hydrophilic and lipophilic fractions of biological samples: Increasing single-injection chemical coverage in untargeted metabolomics. Metabolites. 2021;**11**(5):295

[91] Yaşar Mumin M et al. A new mixed-mode stationary phase derived from [2-(3, 4-epoxycyclohexyl) ethyl] trimethoxysilane as a coupling reagent and its RPLC/HILIC/IEC applications. ChemistrySelect. 2022;**7**(44):e202204069

[92] Wu X et al. In-depth exploration and comparison of chemical constituents from two lilium species through offline two-dimensional liquid chromatography combined with multimode acquisition of high-resolution mass spectrometry. Journal of Chromatography A. 2022;**1670**:462980

[93] Helmer PO et al. Mass spectrometric investigation of cardiolipins and their oxidation products after two-dimensional heart-cut liquid chromatography. Journal of Chromatography A. 2020;**1619**:460918

[94] Dang J et al. Target separation of flavonoids from Saxifraga tangutica using two-dimensional hydrophilic interaction chromatography/reversed-phase liquid chromatography. Journal of Separation Science. 2018;**41**(24):4419-4429

[95] Xu R et al. Lipid wizard: Analysis software for comprehensive two-dimensional liquid chromatography–mass spectrometry-based lipid profiling. Analytical Chemistry. 2024;**96**(14):5375-5383

[96] Zhao R et al. A practical method for amino acid analysis by LC-MS using precolumn derivatization with urea. International Journal of Molecular Sciences. 2023;**24**(8):7332

[97] Hui-ying K et al. Determination of 26 free chiral amino acids in industrial products by pre-column derivatization and HPLC-MS/MS. Journal of Instrumental Analysis. 2024;**43**(5):1-11

[98] Kalmykov PA et al. One-pot determination of amino acids in drugs by pre-column derivatization with phenyl isothiocyanate. Fine Chemical Technologies. 2024;**19**(2):127-138

[99] Pandey R et al. Novel strategy for untargeted chiral metabolomics using liquid chromatography-high resolution tandem mass spectrometry. Analytical Chemistry. 2021;**93**(14):5805-5814

[100] Ardakani MH et al. Sequential purification of cannabidiol by two-dimensional liquid chromatography combined with modeling and simulation of elution profiles. Journal of Chromatography A. 2024;**1717**:464702

[101] Sarrut M et al. Potential and limitations of on-line comprehensive

reversed phase liquid chromatography× supercritical fluid chromatography for the separation of neutral compounds: An approach to separate an aqueous extract of bio-oil. Journal of Chromatography A. 2015;**1402**:124-133

[102] Taylor LT et al. Supercritical fluid chromatography. Analytical Chemistry. 2010;**82**(12):4925-4935

[103] Wei W et al. A high-efficiency strategy integrating offline two-dimensional separation and data post-processing with dereplication: Characterization of bufadienolides in *Venenum Bufonis* as a case study. Journal of Chromatography A. 2019;**1603**:179-189

[104] Jajoo VS et al. Recent advances in supercritical fluid chromatography. Research Journal of Science and Technology. 2024;**16**(1):87-96

[105] Tammekivi E et al. A powerful two-dimensional chromatography method for the non-target analysis of depolymerised lignin. Analytica Chimica Acta. 2024;**1288**:342157

[106] Qu B et al. Combining multidimensional chromatography-mass spectrometry and feature-based molecular networking methods for the systematic characterization of compounds in the supercritical fluid extract of *Tripterygium wilfordii Hook F.* Analyst. 2023;**148**(1):61-73

[107] Teboul E et al. Off-line two-dimensional separation involving supercritical fluid chromatography for the characterization of the wastewater from algae hydrothermal liquefaction. Journal of Chromatography A. 2023;**1694**:463907

[108] François I et al. Comprehensive supercritical fluid chromatography× reversed phase liquid chromatography for the analysis of the fatty acids in fish oil. Journal of Chromatography A. 2009;**1216**(18):4005-4012

[109] Devaux J et al. On-line RPLC x SFC hyphenated to high resolution mass spectrometry for the characterization of third generation bio-oils. Analytics. 2022

[110] Shi X et al. Systematic profiling and comparison of the lipidomes from *Panax ginseng*, *P. quinquefolius*, and *P. notoginseng* by ultrahigh performance supercritical fluid chromatography/high-resolution mass spectrometry and ion mobility-derived collision cross section measurement. Journal of Chromatography A. 2018;**1548**:64-75

[111] Burlet-Parendel M et al. Opportunities and challenges of liquid chromatography coupled to supercritical fluid chromatography. TrAC Trends in Analytical Chemistry. 2021;**144**:116422

[112] Van Beek FT et al. Comprehensive two-dimensional liquid chromatography of heavy oil. Journal of Chromatography A. 2018;**1564**:110-119

Section 2

Specific HPLC Applications

Chapter 5

Utilizing High-Performance Liquid Chromatography (HPLC) in Clinical Diagnostics

Fatemeh Keyfi, Amin Alaei, Hadiseh Mirahmadi Daryasari, Ali Hakimi and Pouya Gharavi

Abstract

High-Performance Liquid Chromatography (HPLC) stands as a robust analytical technique with extensive applications in clinical diagnostics. Its versatility enables the precise separation, identification, and quantification of complex mixtures in biological samples. We propose integrating HPLC into clinical diagnostic procedures to enhance the accuracy, sensitivity, and efficiency of disease detection and monitoring. By leveraging HPLC's capabilities, we aim to streamline the analysis of biomarkers, drugs, metabolites, and toxins in various bodily fluids, facilitating rapid and reliable diagnosis of diseases ranging from metabolic disorders to cancers. Moreover, HPLC's ability to analyze multiple analytes simultaneously can expedite patient assessment and treatment decision-making processes. This chapter advocates for strategically incorporating HPLC technology into clinical diagnostic protocols, ensuring improved patient outcomes through enhanced diagnostic precision and efficiency.

Keywords: vitamins and hormones, inherited metabolic disorders, hemoglobinopathies, pharmacology and toxicology, genetic diagnosis, HPLC, new applications

1. Introduction

HPLC provides highly accurate and reproducible measurements of analytes, which are essential for diagnosing diseases, monitoring therapeutic drug levels, and assessing nutritional status. The technique can detect low concentrations of biomarkers, vitamins, hormones, and drugs in biological fluids specimens such as blood, urine, serum, and tissue extracts, ensuring reliable diagnostic results [1]. HPLC is used to analyze a vast array of compounds, including amino acids, nucleotides, peptides, proteins, vitamins, hormones, and drugs. This versatility makes it indispensable in clinical laboratories [2]. It is an essential tool for quality control in the pharmaceutical industry, ensuring the purity of drugs and the consistency of their production, which directly impacts patient safety [3]. HPLC methods are standardized and validated according to regulatory guidelines (e.g., FDA, EMA), ensuring that diagnostic procedures meet stringent quality and accuracy requirements [4].

HPLC can be coupled with various detectors (e.g., UV, fluorescence, electrochemical, mass spectrometry) to enhance detection capabilities and selectivity based on the nature of the analytes [5]. The technique can be customized with different stationary phases (e.g., C18, ion exchange, size exclusion) and mobile phases (e.g., aqueous, organic solvents) to optimize the separation of specific compounds [6]. HPLC can handle complex biological matrixes, allowing for the separation of analytes from interfering substances. This is crucial for accurate diagnostics in clinical settings where samples are often heterogeneous. Innovations like Ultra-High-Performance Liquid Chromatography (UHPLC) offer faster analysis and higher resolution, while liquid chromatography-tandem mass spectrometry (LC-MS/MS) provides unparalleled specificity and sensitivity for detecting trace levels of biomarkers [7]. HPLC technique is also applicable in Diverse Clinical Areas, including endocrinology for the measurement of hormones, toxicology for detecting and quantifying drugs of abuse and therapeutic drugs, nutrition for monitoring levels of vitamins and minerals to diagnose deficiencies or excesses, hormonology, hematology for hemoglobinopathies detection, and the detection of inherited metabolic disorders [1, 8].

2. Vitamins

Vitamins are organic substances essential for cells to function, grow, and develop properly. HPLC is beneficial for measuring vitamin levels in human blood. Accurate quantification of these vital micronutrients is crucial for diagnosing deficiencies, assessing nutritional status, and guiding treatments [9]. Two types of vitamins are being quantified *via* HPLC: (a) water-soluble vitamins, including B complex vitamins and vitamin C, and (b) fat-soluble vitamins, including vitamins A, D, E, and K. The basis of vitamin analysis through HPLC is similar; however, some differences, such as sample derivatization, the method of detection, or applying mass spectrometry, due to the provision of high sensitivity that is useful for detecting low levels of vitamins and their derivatives and specificity following HPLC, are discussed in the following [10].

2.1 Water-soluble vitamins

Water-soluble vitamins dissolve in water and are quickly eliminated from the body. Since they are not stored in significant amounts, they must be regularly consumed in the diet. The primary water-soluble vitamins include B group vitamins B1 (thiamine), B2 (riboflavin), B3 (niacin), B5 (pantothenic acid), B6 (pyridoxine, pyridoxal, pyridoxamine), B7 (biotin), B9 (folate, folic acid), B12 (cobalamin), and vitamin C (ascorbic acid).

2.2 Fat-soluble vitamins

Fat-soluble vitamins are absorbed alongside dietary fats and stored in the body's fat tissues and liver. They are present in various plant and animal foods as well as in nutritional supplements. The primary fat-soluble vitamins comprise vitamin A (retinol, retinal, and retinoic acid), D (25-Hydroxyvitamin-D3, 25(OH)D3), E (tocopherols and tocotrienols), and K (Phylloquinone (K1), Menaquinones (K2)).

2.3 Sample preparation for vitamins

The preparation of samples for HPLC analysis typically involves three steps: (1) protein precipitation to remove proteins and other interfering substances, (2) extraction, using organic solvents or solid-phase extraction to isolate the vitamins, and (3) derivatization, especially for vitamins that are not naturally fluorescent or UV-absorbing, derivatization can enhance detection sensitivity and specificity [11–14]. Most of the water-soluble and fat-soluble can be derivatized to enhance detection sensitivity, although direct detection methods are also employed. Therefore, common derivatization reagents are not typically used for them, as they are straight forward to analyze directly [15].

2.4 Columns for vitamins

Several types of columns have generally been applied to separate the vitamins. For vitamins, columns are categorized into six types:

(1) Reversed-phase columns, such as C30 (Triacontylsilane, longer alkyl chain that provides more selectivity for structurally similar vitamin D metabolites and enhances the separation of closely related vitamin D metabolites due to increased hydrophobic interactions), C18 (Octadecylsilane), C8 (Octylsilane), columns or similar alkyl-bonded silica that are used as nonpolar stationary phase, and are suitable for separating less polar forms of water-soluble vitamins. These kinds of columns provide high resolution and reproducibility and are compatible with aqueous mobile phases [16, 17], (2) Hydrophilic interaction liquid chromatography (HILIC) columns that contain polar stationary phases, such as silica with polar functional groups (e.g., amino or diol groups), retain highly polar analytes based on hydrophilic interactions. Therefore, they are more efficient for separating highly polar compounds, complementary to reversed-phase chromatography [18], (3) Ion-exchange columns that include Strong Anion-Exchange (SAX) and Strong Cation-Exchange (SCX) columns. These columns separate compounds based on ionic interactions. SAX columns are used for negatively charged water-soluble vitamins, while SCX columns are used for positively charged vitamins [19, 20], (4) Ion-Pair Chromatography Columns are reversed-phase columns with ion-pairing reagents to enhance the retention of ionic compounds by pairing them with counter ions in the mobile phase. This type of column improves peak shape and resolution for ionic analytes [21, 22], (5) Normal-phase columns have been used to separate highly polar water-soluble vitamins. These columns, however, demand precise optimization of the mobile phase and may experience issues with peak tailing. These columns are helpful for differentiating closely eluting compounds that possess aromatic structures [14]. (6) Phenyl Columns, in which phenyl groups provide π-π interactions, offering alternative selectivity for aromatic and planar compounds [23, 24].

2.5 Mobile phases for vitamins

The sort of mobile phases is being used in vitamin HPLC detection are similar. However, some parts of these mobile phases might be diverse. The mobile phase for vitamin B1 contains a mixture of water, methanol, and an acidic modifier (e.g., acetic Acid or formic Acid) [25]. For vitamins B2, B3, B5, B6, B7, B9, C, and K, the mobile phase comprises a combination of water and organic solvents (e.g., methanol or acetonitrile) with an acidic modifier (e.g., acetic acid or formic acid) [20, 26], and the mobile phase for vitamins B12, A, D, and E are as the same as this group with a slight difference in acidic modifier which could be trifluoroacetic acid or formic Acid for B12, acetic Acid for vitamin A, and formic Acid for vitamin D [10, 27].

3. Hormones

Hormones are substances that regulate various functions in the human body by transmitting signals through the human bloodstream to the organs, skin, muscles, and other tissues [28]. Accurate measurement of hormone levels is crucial for diagnosing endocrine disorders, monitoring treatment efficacy, and understanding disease mechanisms [29]. In this section, we examine hormones and the importance of measuring them using HPLC.

Three types of hormones are being quantified *via* HPLC. (1) Steroid Hormones: Including cortisol, aldosterone, and sex hormones (estrogens, androgens, and progesterone) [30], (2) Peptide Hormones: Such as insulin, growth hormone, and parathyroid hormone [31], and (3) Amino Acid-Derived Hormones: Including thyroid hormones (thyroxine and triiodothyronine) [32] and catecholamines (epinephrine, norepinephrine, metanephrine, normetanephrine, and dopamine) [33].

3.1 Steroid hormones

Steroid hormones are a class of hormones that are synthesized from cholesterol and are characterized by their lipid-soluble nature. They can easily pass through cell membranes and typically exert their effects by interacting with specific intracellular receptors [34]. Once inside the cell, they bind to these receptors, forming a hormone-receptor complex that can enter the nucleus and regulate gene expression, ultimately influencing cellular function and activity [35]. Several types of steroid hormones are measured by HPLC, including Cortisol, Aldosterone, Estrogens (Estradiol, Estrone, Estriol), Androgens (Testosterone, Dihydrotestosterone, Dehydroepiandrosterone), and Progesterone.

3.2 Peptide hormones

Peptide hormones are a type of hormone made up of amino acids and are typically water-soluble. They cannot easily pass through cell membranes due to their polar nature and relatively large size. Instead, they attach to particular receptors on the surfaces of target cells, initiating a signaling cascade through second messengers within the cell, ultimately leading to a physiological response [36]. Peptide hormones measured by the HPLC include insulin, growth hormone (GH), and parathyroid hormone (PTH).

3.3 Amino acid-derived hormones

Amino acid-derived hormones, also known as amine hormones, are a class of hormones that are synthesized from single amino acids, typically tyrosine or tryptophan. These hormones are usually small and can have properties similar to both peptide and steroid hormones, depending on their specific structure and function [37]. Amino acid-derived hormones assessed by HPLC include thyroid hormones (thyroxine (T4) and triiodothyronine (T3), thyroid-stimulating hormone (TSH)), and catecholamines (epinephrine (adrenaline) and norepinephrine (noradrenaline) and their metabolites such as metanephrine and normetanephrine, dopamine).

Like vitamins, sample preparation is significant for accurate hormone analysis, although there are some differences. There are two steps for sample preparation in hormone analysis, (1) In extraction hormones are typically extracted from blood using liquid-liquid or solid-phase extractions to be concentrated and purified, and (2) in derivatization, chemical modification is performed on hormones to enhance their detectability. This step is important for those who are not inherently UV-absorbing or fluorescent [38]. Optimal chromatographic conditions, such as column selection, mobile phase, and detection method vary based on the hormone of interest.

3.4 Mobile phases for hormones

The types of mobile phases used in hormone HPLC analysis are generally similar, although some components of these mobile phases might vary, even though they have similarities to the mobile phases for vitamins to some extent. Applying mobile phases for hormones can be classified into three groups of hormones, (1) Mobile phase for steroid hormone detection that contains a mixture of water and organic solvents (e.g., methanol or acetonitrile) with an acidic modifier (e.g., acetic Acid or formic Acid) [39], (2) Mobile phase for peptide hormone analysis involves a gradient elution with mixtures of water, acetonitrile, and trifluoroacetic acid or formic Acid [40], and (3) Mobile phase for amino acid-derived hormone assessment comprises a mixture of water, acetonitrile (or methanol), and an acidic modifier (e.g., acetic acid or trifluoroacetic acid) [41].

3.5 Columns for hormones

Various types of columns are typically used to separate human hormones by HPLC, which are classified into five categories such as (1) Reversed-phase columns, C18 offers high hydrophobicity and is suitable for separating nonpolar to moderately polar hormones such as steroid hormones, C8 is slightly less hydrophobic than C18 columns and can be applied for moderately polar hormones [42], and phenyl columns are applicable for separating hormones that have aromatic structures [41], (2) Normal-Phase Columns, silica columns, which are used for separating highly polar hormones that are not well retained on reversed-phase columns [43], (3) HILIC columns, which are employed for high-polar hormones that are difficult to retain on reversed-phase columns. HILIC columns utilize a mobile phase rich in water and a stationary phase that is hydrophilic, making them effective in separating hormones [44], (4) Ion-Exchange Columns are used for separating charged hormones or hormone metabolites, such as thyroid hormones and their derivatives as well as catecholamines [45], and (5) Size-Exclusion Columns that are particularly useful for peptide and protein hormones [46].

4. Diagnosis of inherited metabolic disorders

4.1 Inherited metabolic disorders (IMDs)

Inborn errors of metabolism (IEM) were first described by A- Garrod in 1908, indicating diseases due to deficiency in a metabolic pathway caused by the faulty activity of a specific enzyme [47]. IMDs encompass more than 200 single-gene disorders that are inherited in an autosomal recessive manner. Many of these diseases carry significant clinical outcomes to the affected newborn children, which include irreversible mental deferment, physical handicaps, or even death. IEM can be arranged into two main categories: those that can influence energy production and those that affect the synthesis or degradation of specific molecules [48]. A fault in the metabolic pathway can result in the utilization of another fuel source. This change causes increases or decreases in abnormal metabolites in patients with these disorders, which is critical to diagnosis. Organic acidemias due to mitochondrial enzyme deficiency in the catabolism of branched-chain amino acids (AAs) pathway and fault in fatty acid oxidation pathways made a group of >20 diseases in which acyl-Coenzyme A (acyl-CoA) esters increase in the mitochondria [2].

Precise diagnosis of IMD is crucial for obtaining timely and relevant patient outcomes. Analyzing amino acids in physiological matrixes is employed to diagnose these disorders. Plasma amino acids can be examined using RP-HPLC [49]. Additionally, the serum acylcarnitine profile can be readily identified using LC-MS/MS. However, profiling acylcarnitines in urine is more complex. Various techniques, including column chromatography, HPLC, and GC/MS, can be employed. Several HPLC techniques have been developed for the analysis of amino acids, offering shorter analysis times [50]. This is particularly important when screening for metabolic disorders. For instance, the diagnosis and therapy monitoring of infants with inborn errors in homocysteine metabolism involves the quantification of plasma methionine. Allo-isoleucine, an isomer of leucine and isoleucine found in patients with maple syrup urine disease (MSUD), can be separated using conventional HPLC methods [51].

4.1.1 LC-MS/MS utilization for IEM profile

Coupling HPLC with MS offers significant benefits in selectivity and sensitivity. The initial adoption of LC-MS/MS was driven by its success in quickly analyzing plasma and urine acylcarnitine profiles (2–3 minutes) [52]. These profiles have proven particularly diagnostic for the recently identified fatty acid oxidation defects (FOADs) and for organic acidemias stemming from disorders in branched-chain amino acid catabolism [53]. Blood spots are used instead of plasma because plasma contains very few long-chain acylcarnitines (C12 to C18). This method is employed for the analysis of neutral, acidic, and essential amino acids, as well as for the diagnosis and screening of phenylketonuria (PKU), MSUD, tyrosinemia, citrullinemia, and acylcarnitine profiles [54].

4.2 HPLC utilization for PKU diagnosis

Phenylketonuria (PKU) is an autosomal recessive metabolic disorder caused by mutations in the phenylalanine hydroxylase (PAH) gene. These mutations render PAH unable to convert phenylalanine (Phe) into tyrosine (Tyr), leading to elevated levels of Phe in the blood [55]. If untreated, PKU can result in detrimental levels of Phe in the human body, resulting in cognitive impairment and other significant health problems. Therefore, an effective screening test is essential and should be conducted as early as possible. In developed countries, most newborns are screened for PKU shortly after birth [56]. Measuring Phe levels in the blood is crucial for diagnosing PKU and managing the condition through diet [57].

Furthermore, detecting blood Phe and Tyr concentrations simultaneously and calculating the Phe/Tyr ratio can lower the false-positive rate in PKU screening. HPLC is a highly valuable method for determining Phe and Tyr levels [58]. Several HPLC methods are utilized for the precise quantification of Phe and Tyr in serum, plasma, and dried blood-spot samples. Nevertheless, these techniques have drawbacks, such as the need for complex derivatizing agents or intricate sample preparation procedures [59]. For instance, phenylisothiocyanate reagent has been employed for amino acid pre-column derivatization. The HPLC method with ultraviolet detection (HPLC-UV) is a straightforward, fast, precise, convenient, and cost-effective approach. It proves highly beneficial for screening, diagnosing PKU, and monitoring dietary therapy [60].

4.3 Amino acids disorders diagnosis with HPLC

The level of total amino acids, such as individual amino acid levels in biological fluids, has become an essential practice in clinical laboratories. The total amino acids percentage in plasma and urine indicates the nutritional/metabolic situation. The person-specific amino acid composition of biological fluids is important in verifying or otherwise ruling out the suspected amino acidopathies [61].

Amino acid analysis in physiological fluids is widely employed for diagnosing and screening inherited metabolic disorders. The primary technique, ion-exchange chromatography (IEC) with ninhydrin identification, is preferred due to its capability to analyze all clinically significant amino acids [62] in a single assay, offering precision and requiring minimal sample preparation. Various techniques using HPLC have also been developed for amino acid analysis [63]. These methods provide quicker analysis times but may not separate all amino acids that can be detected using IEC [62]. This is particularly useful in metabolic disorder screening. Elevated levels of valine, leucine, and isoleucine found in maple syrup urine disease (MSUD) patients, which are characteristic of this condition, can be distinguished from each other using traditional HPLC techniques. Additionally, allo-isoleucine can be separated and precisely measured *via* HPLC [64]. **Figure 1** provides a comprehensive comparison of the amino acid profile obtained through HPLC analysis.

4.4 HPLC method for the simultaneous identification of organic acids

Organic acid disorders are a diverse set of inherited metabolic conditions where organic acids accumulate within the body. These disorders are notably prevalent in regions with elevated rates of consanguineous marriages [65]. To assess patients with organic acidemias and various other IEMs, methods such as LC-MS/MS, gas chromatography-mass spectrometry (GC/MS), and isocratic cation-exchange HPLC can be employed [66].

HPLC method for the concurrent identification of organic acids uses a single wavelength of UV detection. UV can detect organic acids by detecting absorption *via* their carboxyl groups [67]. Moreover, employing a low-carbon octadecyl silyl (ODS) column decreased the hydrophobic nature of the silica surface, ensuring a reliable method for separating highly polar compounds like organic acids in a 100% aqueous solution [68].

An HPLC method has been developed to concurrently measure urinary organic acids and creatinine as part of studying aromatic solvent metabolism. Following acidified urine extraction with diethyl ether and subsequent evaporation, the dried residue is reconstituted in the mobile phase, with hydroxybenzoic acid serving as an internal

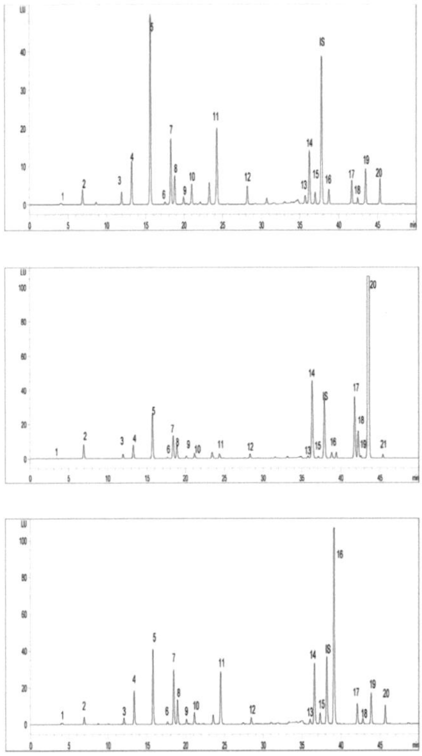

	Amino Acids
1	Aspartic Acid
2	Glutamic Acid
3	Aspargine
4	Serine
5	Glutamine
6	Histidine
7	Glycine
8	Threonine
9	Citruline
10	Arginine
11	Alanine
12	Tyrosine
13	Methionine
14	Valine
15	Tryptophane
16	Phenylalanine
17	Isoleucine
18	Allo- Isoleucine
19	Ornithine
20	Leucine
21	Lysine

Figure 1.
Comparison of amino acid profile with HPLC (A: normal patient, B: MSUD patient, C: PKU patient).

standard [69]. A Nucleosil C18 column is utilized along with a pre-column of identical material. The mobile phase consists of acetonitrile and phosphate buffer at pH 3.3 in a ratio of 17:83. For creatinine determination, the sample is diluted tenfold and the eluate is monitored at 215 nm using UV detection. This method provides consistent results and is straightforward, dependable, and valuable for biological monitoring [70].

4.5 Metabonomics

It is specified as the "quantitative measurement of the dynamic multi-parametric metabolic responses of living systems to pathophysiological motive or genetic adjustment" [71]. It is employed to characterize the biochemical profiles of naturally occurring metabolites within cells, bodily fluids, or tissues, enabling physiological evaluation, disease diagnosis, and prognosis. Unlike traditional biochemical methods that typically target individual metabolites, metabonomics examines a spectrum of molecules encompassing various small molecules like peptides, amino acids, organic acids, lipids, nucleic acids, sterols, carbohydrates, and vitamins. Thus, it offers a comprehensive view of how pathophysiological processes or pharmacological treatments influence metabolism and metabolic dynamics [72].

The UPLC technique is valuable for metabonomics analysis caused of its heightened sensitivity in detecting metabolites, which is well-suited for extensive untargeted investigations. Operating with 1.7 μm chromatographic particles and a fluid system capable of handling pressures between 6000 and 15,000 psi, UPLC offers enhanced chromatographic selectivity compared to HPLC, which employs larger particles.

The narrower peak widths achieved with UPLC result in improved signal-to-noise ratios and increased sensitivity compared to conventional HPLC. This enables better resolution of peaks and faster separation of complex mixtures. Additionally, UPLC's superior resolution significantly mitigates issues related to ion suppression [73].

4.5.1 Type 1 diabetic (T1D)

Insulin is a primary hormone released after meals, exerting a significant influence on carbohydrate, lipid, and protein metabolism. Serum metabonomics shows specific metabolic disorders before β-cell autoimmunity in humans. It can be applied to the treatment of T1D children [74]. This research described different pathways associated with metabolism as alternative targets for therapy to diabetic patient treatment. Another UPLC-based metabonomic result showed significant distribution in plasma amino acids and even amino acid derivatives during insulin exclusion in T1D due to multiple metabolic pathways disturbed in acute insulin deprivation T1D. Furthermore, in untreated T1D, because of the breakdown of muscle protein and the release of amino acids in the circulation and liver, plasma branched-chain amino acids are increased [75].

4.5.2 Type 2 diabetes (T2DM)

T2DM and its associated complications represent a major global public health concern. T2DM is a common metabolic disorder marked by insulin resistance and a relative insufficiency in insulin production [76]. The metabolism of fatty acids and levels of free fatty acids undergo specific changes in diabetic patients. Elevated plasma acylcarnitines and tryptophan levels, along with decreased levels of LPC (16:0), LPC (18:0), LPC (18:2), and phenylalanine, have been observed in treated individuals with type 2 diabetes mellitus (T2DM). There are also reported alterations in plasma amino acids and their distribution across metabolic pathways involving 3-indoxyl sulfate, glycerophospholipids, and free fatty acids, as well as potential impacts on bile acids in diabetic patients. Fatty acids enhance insulin secretion under basal or glucose-stimulated conditions and are crucial for stimulus-secretion coupling in β-cells. However, prolonged elevation of free fatty acid levels can lead to or worsen insulin resistance, contributing to the onset and progression of type 2 diabetes. Elevated circulating free fatty acids may inhibit insulin receptor substrate (IRS) function over time [77].

5. Hemoglobinopathies and HPLC applications

Hemoglobinopathy, thalassemia, and porphyria are vital diseases of hemoglobin that affect the structure and biosynthesis of heme. There are various methods to identify hemoglobin-related disorders, each depending on the particular disorder. Due to HPLC's high precision and sensitivity, this technique is widely used in identifying and managing these disorders. This section will discuss the different types of hemoglobin disorders, their underlying causes, and the HPLC technique's application in their identification.

5.1 Introduction to hemoglobinopathy

Hemoglobinopathy is related to genetic disorders of hemoglobin and is generally divided into two categories: thalassemia syndrome and abnormal structures of hemoglobins [78]. Abnormal hemoglobins result from genetic mutations in the genes responsible for producing globin chains. These mutations can alter the physical and chemical properties of the hemoglobins, leading to various hemoglobinopathy diseases such as Hemoglobin S (HbS), Hemoglobin C (HbC), Hemoglobin E (HbE), and Hemoglobin D (HbD).

5.2 Thalassemia

Thalassemia encompasses a class of genetic blood diseases characterized by insufficient production of hemoglobin, leading to anemia. Thalassemia is primarily categorized into alpha and beta types, based on which specific part of the hemoglobin molecule is affected.

5.2.1 Alpha thalassemia

Alpha thalassemia is caused by mutations or deletions in the HBA1 and HBA2 genes located on chromosome 16, which are essential for producing alpha globin chains in hemoglobin. Typically, an individual possesses four alpha globin genes, with two inherited from each parent. The clinical severity of alpha thalassemia correlates with the number of these genes that are altered. Types of alpha thalassemia are Silent carrier state, Alpha thalassemia trait (Minor), Alpha Thalassemia Major (Hydrops Fetalis) [79], and Hemoglobin H disease [80].

5.2.2 Beta thalassemia

Beta thalassemia arises from mutations in the HBB gene on chromosome 11, impacting the production of beta-globin chains in hemoglobin. These mutations can either completely inhibit beta-globin production ($\beta 0$ mutations) or partially reduce it ($\beta +$ mutations). Types of beta thalassemia include Beta Thalassemia Minor (Trait), Beta Thalassemia Intermediate, and Beta Thalassemia Major (Cooley's Anemia) [81]. There are also other forms of thalassemia, such as Hemoglobin E (HbE) and sickle cell beta-thalassemia (S/β).

5.3 The HPLC role in the hemoglobinopathies and thalassemia diagnosis

Ion-exchange high-performance liquid chromatography, especially cation-exchange HPLC (CE-HPLC), is a commonly utilized laboratory test for detecting hemoglobin types and quantifying hemoglobin F and A2 for the thalassemia diagnosis. CE-HPLC has become popular due to its speed, versatility, reproducibility, and convenience as the screening technique of selection for hemoglobin methods [82].

For the measurement and detection of hemoglobin by HPLC, fresh whole blood is collected in tubes containing EDTA and used as samples [83].

In the study that was conducted in India, HPLC was used to detect and measure the prevalence of hemoglobin variants and thalassemia. In this research, a total of 65,779 cases were examined for hemoglobinopathies using the Bio-Rad Variant HPLC system with the beta-thalassemia brief program. They identified the types of hemoglobin and detected hemoglobinopathies based on retention times, the proportion of hemoglobin (%), and peak features. Additionally, using thalassemia patterns, they successfully identified the specific types of thalassemia [83].

In another clinicohematological study involving 110 patients with thalassemia and hemoglobinopathy, CE-HPLC was used for evaluation. Among the 110 cases, 87 were diagnosed with thalassemia disorders and 23 with hemoglobinopathy disorders. Additionally, various types of thalassemia were identified using the thalassemia profile. The study identified various types of thalassemias among the participants, such as β-thalassemia trait, thalassemia intermedia, thalassemia major, HbH disorder, compound heterozygotes, HbS/β thalassemia, and HbE/β-thalassemia, using the CE-HPLC technique. Hemoglobinopathy cases detected by CE-HPLC include HbE trait, HbD Punjab trait, HbS trait, HbD Iran, HbDD, HbSS, HbEE, HbJ Oxford, and compound heterozygotes for HbS and HbE [84]. These findings illustrate the effectiveness of the HPLC technique in diagnosing and differentiating between various hemoglobinopathies and thalassemias.

6. Porphyria: A disorder in the hemoglobin synthesis pathway

Porphyria includes a set of rare metabolic disturbances resulting from defects in the heme biosynthesis pathway. These disorders lead to the buildup of porphyrins or their precursors, leading to various symptoms, including photosensitivity, abdominal pain, and neuropsychiatric manifestations. These disorders cause an overproduction of intermediates in the biosynthesis pathway of heme. The increased production and accumulation of these intermediates in the blood lead to clinical manifestations. Additionally, the heightened production results in increased excretion through urine and feces. Porphyrias remain underdiagnosed despite the potential for laboratory tests measuring metabolites to confirm clinical suspicion in symptomatic patients. Additionally, assessing enzymatic activities and conducting molecular analyses are essential for confirming the diagnosis and identifying pre-symptomatic carriers [85].

6.1 The use of HPLC in diagnosing porphyria

Porphyrins are measured using various methods, including photometric techniques, enzyme activity assays, and molecular studies. One of the most widely used techniques for investigating this disease is HPLC, which will be discussed further. Samples used in the investigation of these disorders include serum, urine, and stool. In situations where acute porphyria has developed, and urine analysis is not feasible, such as in cases of kidney failure, measuring PBG in serum can be beneficial. In acute porphyrias, the levels of PBG in serum are elevated, though they are lower than in urine when kidney function is normal. As advances in the pathway of heme biosynthesis, the increasing hydrophobicity gradient causes the more hydrophilic metabolites (ALA, PBG, URO, COPRO) to be predominantly excreted in the urine, while the relatively hydrophobic metabolites (COPRO, PROTO) are primarily excreted in the stool [85].

6.1.1 Measurement of porphyria in urine

Multiple methods exist to evaluate porphyria, but RP-HPLC with fluorescence detection (FLD-HPLC) is regarded as the gold standard for measuring porphyria disorder [86]. The ability of this method to differentiate diverse kinds of porphyrias and their isomers has established it as the premier method for porphyria diagnosis [87].

Typically, a C-18 column is used as the stationary phase for chromatographing urine porphyrins, while a linear gradient elution system is employed for the liquid phase. In this system, phase A consists of 10% acetonitrile by volume in 1 M ammonium acetate, and phase B consists of 10% acetonitrile by volume in methanol. Porphyrins with two to eight carboxylic groups, including the differentiation between type I and type III isomers, can be separated in less than 30 minutes. The excitation and emission wavelength ranges typically applied are approximately 395–420 nm and 580–620 nm, sequentially. Since a random urine sample is used to evaluate porphyria, the results from HPLC analysis are normalized to the urine's creatinine concentration, with the final results expressed in nanomoles per millimole of creatinine.

The concentration of excreting porphyrins in urine in healthy people is less than 35 nmol/mmol creatinine. Additionally, in their urine, coproporphyrin (COPRO) is excreted more than uroporphyrin (URO), with COPRO isomer 3 being more prevalent than isomer 1. Hepacarboxyl, hexacarboxyl, and pentacarboxyl porphyrins are excreted in minimal amounts. In porphyric patients, the urinary excretion of porphyrins differs depending on the specific enzymatic deficiency associated with the specific porphyria type and its phase of the disease [85].

The initial laboratory test for detecting acute porphyria in patients experiencing acute neurovisceral attacks is the assessment of ALA and PBG levels [88]. ALA and PBG concentrations in urine specimens are evaluated using ion-exchange HPLC techniques. Both anion-exchange and cation-exchange columns are employed to measure these metabolites. In summary, urine specimens are processed through two sequential columns. The top column, which contains an anion-exchange resin, adsorbs PBG. ALA, which passes through this top column, is then retained by the cation-exchange column positioned below [85].

6.1.2 Measurement of porphyria in feces

Another sample that is used to diagnose porphyria is feace. Analyzing the metabolites and their patterns in feces can assist in identifying the type of porphyria. For this analysis, 25–50 mg of feces are processed by sequentially adding 1 ml of concentrated hydrochloric Acid to dissolve the organic matrix, 3 ml of diethyl ether to remove contaminants such as chlorophylls and carotenoid pigments, and 3 ml of water to prevent any changes in protoporphyrin. By applying the protocol used for diagnosing and evaluating heme synthesis intermediates in urine, we can also assess the prepared stool sample. Analyzing the patterns of excreted metabolites in feces in porphyria patients reveals the presence of enzyme disorders associated with the disease's progression [85].

7. HPLC applications in pharmacology and toxicology

HPLC's Recent Advancements are Improved detection technology (like mass spectrometry) and advanced sample preparation techniques (like bioaffinity chromatography online automation). These advancements enhance the capabilities of HPLC and ensure its continued use as a gold standard for clinical testing of biomarkers and therapeutic drugs [89].

HPLC fulfills the need for sensitive, timely, and reliable analytical results in clinical chemistry. Over the past decade, HPLC has largely replaced spectroscopic methods and gas chromatography in the pre-sale testing, marketing, and regulation of drugs. Initially seen as a complementary technique to gas chromatography, HPLC has now become the preferred method for pharmaceutical analysis [3].

Several factors, including the polarity of the liquid mobile phase, a wide range of stationary phases, and a specific and sensitive detector system, contribute to the widespread and effective use of HPLC in pharmaceutical analysis. Challenges in drug analysis using HPLC are Low analyte concentrations, Structural similarity of drugs and endogenous compounds, Drug binding to plasma proteins, Extraction techniques to isolate drugs from complex body fluids, and separation systems to resolve drugs from co-extractives. HPLC's versatility and sensitivity make it a valuable tool for drug analysis, enabling accurate and reliable determination of drug concentrations and aiding in clinical research and drug development [3].

HPLC can face limitations due to its potential lack of specificity. Certain analytes may exhibit weak UV chromophores, making UV detection challenging. Additionally, some components may be strongly retained on the HPLC column. UV detection offers advantages in terms of linearity and enables rapid quantitative analysis using a single standard. Diode array and rapid scanning detectors assist in identifying peaks and monitoring their purity but have lower sensitivity compared to single wavelength detectors. Fluorescence and electrochemical detectors provide enhanced sensitivity and selectivity for specific analytes, surpassing UV detectors in many cases. Electrochemical detection, in particular, exhibits high sensitivity and has yielded promising results in investigating certain drug classes. Ion-exchange chromatography on unmodified silica columns remains a common technique for analyzing basic drugs. Reversed-phase systems work well for neutral or weakly acidic drugs. Acidic drugs, including paracetamol and cannabis, can be separated using ion suppression or ion-pair chromatography techniques on reversed-phase packing materials. For basic drugs, ion-exchange chromatography is a widely used technique. This method uses silica columns without modification with an eluent buffered to a pH of around nine. For neutral and weakly acidic drugs like barbiturates, reversed-phase systems are commonly used.

When compounds exist in different isomeric forms, it is often necessary to separate them by employing liquid chromatography. This is particularly crucial for pharmaceutical compounds, as isomers can exhibit different toxicological and pharmacological properties. Enantiomers present a unique challenge as they possess identical physicochemical characteristics and cannot be distinguished using conventional chromatography. Therefore, separating enantiomers in liquid chromatography requires a "chiral selector." This can be achieved through a chiral derivatization agent, a chiral mobile phase additive, or a chiral stationary phase. Diastereoisomers, which possess two "chiral centers," display differing physicochemical characteristics, enabling their separation using standard "achiral" liquid chromatography columns. Nonetheless, using a chiral stationary phase is usually favored over employing a chiral derivatization agent for several reasons. Employing a chiral stationary phase is often preferred over a chiral mobile phase additive, in part because using a chiral selector as a mobile phase additive results in a greater application of a frequently costly substance. Enantiomerically pure chiral compounds are naturally occurring and readily accessible at a relatively low cost. These substances serve as the foundation for most commercially accessible chiral stationary phases. The most frequently used chiral stationary phases are derived from proteins, cyclodextrins, derivatized polysaccharides, and derivatized amino acids.

Liquid chromatography plays a critical role in drug dissolution studies, which evaluate the potential availability of a drug substance from a pharmaceutical formulation upon entering the stomach. The mixture is agitated in a bath to dissolve in an aqueous buffer, mimicking the stomach condition. Samples of the buffer are collected at set time intervals and analyzed to determine the drug concentration. Monitoring the dissolution of excipients is also beneficial, as can affect how the drug is released. Liquid chromatography can be used to study this, but it faces challenges due to the weak chromophores or polymeric nature of the excipients.

Stability studies in pharmaceuticals are crucial to prevent the formation of potentially toxic degradation products. These studies aim to demonstrate the stability of the drug content in a formulation over time. Additionally, if degradation occurs, it is essential to identify and quantify the degradation products. A notable example is the development of liquid chromatography conditions for determining pilocarpine in ophthalmic solutions, which can also be used to analyze its degradation products, isopilocarpine, and pilocarpic Acid. Stability studies not only focus on the degradation of the active ingredient but also consider the possible influence of excipient degradation on the drug release characteristics of the formulation. In many pharmaceutical quality control laboratories, microbiological assays have traditionally been the official methods for antibiotic analysis. However, HPLC is increasingly employed for the quantitative determination of antibiotics, offering significant advantages in the field of drug quality control [3].

Developing and validating analytical methods are essential parts of pharmaceutical development programs. HPLC analysis methods are designed to identify, quantify, or purify substances of interest. HPLC plays a significant role in studies assessing the stability of drug formulations, ensuring the stability and efficacy of drugs over their shelf life. It has been successfully employed in stability studies of various drugs, including atropine, antibiotics, and biotechnology-based drugs such as insulin and streptokinase [89].

HPLC is a vital tool in clinical research, enabling the analysis of diverse biological samples due to its versatility which makes it suitable for a broad array of clinical uses, leading to advancements in healthcare and drug development, including [90, 91] separation of phenyl alkylamines and detection of poisons/intoxicants and illicit drugs in urine and blood, forensic toxicology, analysis of pesticide content in drinking water, assay and quality control of cosmetics, separation of plant components with similar structures (such as cinchona, digitalis, ergot extracts, licorice), separation and analysis of psychotropic drugs, drug stability testing, Analysis of drug degradation products (such as stability studies of atropine), determining the activity of compounds like chloramphenicol and antibiotics, monitoring antibiotic production (e.g., chloramphenicol, tetracyclines, streptomycins), monitoring synthesis reactions and isolating products, evaluating biopharmaceutical properties and pharmacokinetics of drugs, analytical method for various natural and synthetic drugs, investigating the absorption, distribution, metabolism, and excretion of drugs in the human body, identifying, quantifying, and characterizing drugs and their metabolites, providing details regarding the composition of samples related to drugs.

7.1 HPLC applications in monitoring immunosuppressor drugs

Immunosuppressant drugs, including cyclosporine, sirolimus, and tacrolimus, have significantly improved the survival rates of transplant patients. However, due to their complex mechanisms of action, variable absorption rates, and narrow therapeutic indices, patients require close monitoring during treatment to avoid adverse drug reactions and prevent organ transplant rejection. The primary methods used for immunosuppressant drug monitoring are HPLC and immunoassays. A major limitation of immunoassays is their potential for cross-reactivity with metabolites, even when using monoclonal antibodies. On the other hand, HPLC-based methods have remained the gold standard for therapeutic drug monitoring of immunosuppressants due to their unmatched specificity, sensitivity, and multiplexing capabilities, allowing for quantifying multiple drugs and their metabolites simultaneously from a single sample. HPLC with UV or mass spectrometry detection is routinely employed for therapeutic monitoring of immunosuppressants [92].

7.2 HPLC applications in monitoring alcohol abuse

The growing problem of chronic alcohol abuse has made it crucial to effectively monitor alcohol consumption. Carbohydrate-deficient transferrin (CDT) has become a specific marker for excessive alcohol use. Individuals with a history of heavy drinking over an extended period typically exhibit elevated levels of CDT glycoforms. A key advantage of CDT over traditional biochemical markers of chronic alcohol abuse, like liver enzymes, is its high specificity, leading to fewer false-positive results. Various analytical methods have been developed for CDT analysis. Isoelectric focusing (IEF) was initially employed to separate, detect, and quantify transferrin glycoforms. Glycoforms with an isoelectric point at pH 5.7 or above, after complete iron saturation, are collectively known as CDT. Several commercial and noncommercial methods are available for routine CDT determination, including ion-exchange chromatography on minicolumns followed by immunoassay (e.g., the Bio-Rad % CDT-TIA) and HPLC. Among these methods, HPLC is the preferred choice due to its ability to provide reproducible separation and visible detection of different transferrin glycoforms. This is particularly valuable because of genetic transferrin variants and glycoform kinds that can lead to inaccurate CDT determination using minicolumn immunoassays can be readily identified by their unique peak patterns in the chromatogram [93].

HPLC is an essential tool in the pharmaceutical industry, playing a vital role in identifying and characterizing new drug candidates, optimizing formulations, ensuring drug purity, analyzing pharmaceutical substances, purifying compounds, conducting trace analysis, and ensuring food safety by detecting pesticides and toxic chemicals in food products. HPLC's versatility, accuracy, and efficiency make it a preferred technique for routine and rapid analysis in these fields [90].

8. Applications of HPLC in genetic diagnosis

8.1 Using HPLC in cancer diagnosis

During the development of a tumor from a uniform proliferating clone to a diverse group of cell subpopulations, cells produce a wide range of enzymes and surface molecules. These molecules, along with those generated by the host as a reaction to the tumor, are collectively referred to as tumor markers. Additionally, biomarkers can be utilized for early detection and identification of potential premalignant lesions, predicting disease prognosis, monitoring therapy, and prompt identification of recurrence or developing a second primary cancer [94].

8.1.1 HPLC analysis of the protein profile in cervical cancer

Cervical cancer ranks as the third most prevalent cancer globally and is the second most common cancer in women, following breast cancer. In developing countries, it is the primary cause of cancer-related deaths among women. Understanding the molecular origins and interactions that lead to malignancy can assist in early diagnosis. Squamous cell carcinoma is the most prevalent type of malignant cervical tumor, with oncogenic human papillomavirus types 16, 18, 30, and 31 identified as causative agents. Numerous biochemical changes occur in the body during the induction of malignancy, and by the time a tumor is first detected, it may accommodate a billion cells or more. Cancers develop slowly, sometimes over several years. Effective screening initiatives along with early diagnosis and treatment, can lower mortality rates. Pap smear screenings detect premalignant lesions and early subclinical diseases, facilitating effective treatment and cure of cancer. However, false negative results are recurrent, and despite screening, significant incidences of cervical cancer persist in screened populations. Abnormal Pap smears are usually assessed using colposcopy, a more precise but costly diagnostic method that necessitates skilled personnel. Emerging optical spectroscopic methods, such as fluorescence and Raman spectroscopy, show promise for early malignancy diagnosis. The use of tumor markers for early diagnosis through proteomics methods is also gaining traction. Diagnosing tumor markers with HPLC plays a crucial role in early cancer detection. These extremely sensitive, precise, and objective techniques offer insights into the structural changes and molecular composition of biochemical entities, which is invaluable for early detection. Optical identification can be performed in real time by trained technicians, potentially diminishing the need for clinical expertise, biopsies, and subsequent pathological examination. HPLC is an effective separation method that can detect individual elements in intricate biological fluid mixtures. When paired with a very sensitive detection technique such as laser-induced fluorescence, it can separate and detect many proteins present in biological samples in a single run at sub-femtomole levels [95].

8.1.2 Using HPLC for early detection of oral cancer

Oral cancer is one type of cancer for which few markers have been identified so far. Possible markers include squamous cell antigens (glycoproteins, 44–48 kDa), epithelial keratins, immunosuppressive cytokines, and cell surface antigens. In normal saliva, the major proteins are mucin, α-amylase, and lipase, with trace amounts of immunoglobulins. Unfortunately, no singular marker specific to oral cancer has been found that is not present in normal or benign lesions. Combining multiple markers detected simultaneously is highly likely to enhance the success rate of early cancer detection. By combining ultrasensitive optical techniques such as laser-induced fluorescence with highly effective separation methods like HPLC, we can detect minute quantities of individual biomolecules in intricate, multicomponent physiological systems. HPLC-LIF, along with the spectroscopy of saliva and serum, can provide spectra and chromatograms of multicomponent physiological samples to detect tumor markers. Unlike immunoassays, which are limited to detecting or estimating one known marker at a time, HPLC can analyze multiple markers in a single run, thereby enhancing diagnostic capacity [94].

8.1.3 Diagnosis of breast cancer using HPLC

Breast cancer represents a major health risk for women worldwide, leading to substantial mortality and morbidity. Various studies have explored the analysis or profiling of metabolite target methods focusing on nucleosides for breast cancer diagnosis. However, these approaches necessitate precise quantification of nucleoside and creatinine concentrations in samples, and their diagnostic efficacy relies on the number of nucleosides measured. The complex nature of their instrument analysis procedures renders them impractical for routine clinical use. Currently, HPLC metabonomics fingerprinting is advocated as a promising diagnostic approach. A computational method has been proposed for analyzing human urine fingerprints to distinguish between individuals with breast cancer and those who are healthy [96].

The use of tumor markers for early detection through numerous proteomics-based techniques is gaining significance due to their high sensitivity, specificity, and objectivity. Analyzing changes in tissue protein profiles offers specific advantages by providing insights into structural changes and molecular composition, enhancing understanding of disease progression. Recent advancements in separation and detection techniques, such as nano-HPLC and multidimensional HPLC for protein and peptide separation, have fueled dynamic growth in proteomics. HPLC, an efficient separation method, is capable of isolating and identifying individual components in complex biological fluids like serum and tissue homogenates. Coupled with compassionate identification methods such as laser-induced fluorescence, HPLC can effectively separate and detect proteins in biological samples at sub-femtomolar levels [97].

8.1.4 Colon cancer detection by HPLC

Despite advancements, early detection of colon cancer remains challenging, and current methods, including monoclonal antibodies (MAb), often fall short. Screening programs such as sigmoidoscopy, colonoscopy, and barium enemas do not consistently yield positive results. The prospect of a serologic test for early colon tumor detection is promising, potentially identifying high-risk individuals or those already with malignancies. While immunochemical methods are extensively utilized for diagnosis, findings are not consistently definitive. The use of gel fiberglass (GFG) embedded with receptors has introduced a new type of affinity matrixes for isolating proteins, including tumor-associated antigens (TAA). Among these TAAs, the p53 protein is significant. It exists in two forms within cells: nuclear and cytoplasmic. While the nuclear form is extensively studied for its association with chromosomal aberrations and gene mutations in colon cancer, less attention has been given to the cytoplasmic, soluble p53 protein in spite of its diagnostic relevance. The p53 protein is considered a crucial cancer marker primarily detectable through histochemical methods, with serological methods also proving successful. Using ELISA and commercially available recombinant p53 antigens, scientists have identified p53 antibody levels in the blood of cancer patients. A significant correlation has been noted in colon cancer between serum p53 antibody levels and prognosis [98].

8.1.5 Detecting liver cancer through HPLC

Metabonomics involves investigating, profiling, and fingerprinting metabolites in different physiological states. Recently, this technique has been applied to diagnose coronary heart disease rapidly and non-invasively. Traditionally, metabolite profiling focuses on analyzing specific classes of metabolites, but metabonomics aims to encompass all compound types and uses metabolic fingerprinting to swiftly categorize samples based on their origin and biological significance. Achieving a stable metabolite fingerprint is crucial for optimizing and utilizing metabonomics effectively.

Unlike traditional metabolite target analysis, which focuses on a predefined set of metabolites, metabonomics utilizes technologies like GC-MS, nuclear magnetic resonance (NMR), and LC-MS to generate data. Recent work by Tuan et al. demonstrated the suitability of HPLC in metabonomics, offering advantages in experimental control, data acquisition, and analysis optimization.

A significant application of metabonomics is in studying nucleosides, an important class of metabolites that can potentially serve as tumor markers. However, challenges include false positives in benign and inflammatory diseases, complicating nucleoside analysis as cancer biomarkers. There are approximately 93 modified nucleoside metabolites mainly derived from tRNA, further complicating their utility in cancer detection methods, particularly in distinguishing liver cancer from inflammatory conditions like hepatitis.

HPLC-based metabonomics has shown promise in distinguishing between cancer patients and healthy individuals, as well as among patients with liver cirrhosis or hepatitis and individuals with liver cancer. These findings highlight the potential of metabonomics in detecting new clinical markers for cancer and other disorders in the post-genomics era. Complementing NMR, HPLC-based metabonomics offers advantages such as simplicity and cost-efficiency [99].

8.1.6 Detection of human bladder cancer using HPLC

Bladder cancer ranks as the second most prevalent genitourinary cancer. Many patients undergo recurrence but generally do not succumb to the disorder. Although bladder cancer is the fourth most common cancer in men, following lung, prostate, and colorectal cancers, it stands as the second most frequently diagnosed malignancy among middle-aged and elderly men after prostate cancer.

Current standard methods for detecting and monitoring bladder cancer include cystoscopy, voided urine cytology, and imaging. Nevertheless, cystoscopy is invasive, uncomfortable, and expensive, making it unsuitable as a screening test. Urine cytology, though noninvasive, has limited sensitivity (20–40%) for detecting low-grade tumors. Various techniques have been explored for early bladder cancer detection through potential biomarkers, but their specificity and sensitivity range widely (50–100%), rendering them inadequate for screening.

Preferably, a urine-based marker for bladder tumors should be noninvasive, cost-effective, easy to use, and highly accurate. Such ideal markers would function for multiple purposes including screening, initial diagnosis, monitoring recurrence and progression, and predicting prognosis. HPLC-MS can detect metabolite variations that distinguish urine specimens from healthy individuals and those with bladder cancer [100].

8.2 Utilizing denaturing HPLC for SNP identification and genotyping

Denaturing HPLC (DHPLC) separates polymerase chain reaction (PCR) amplified products in a specialized column using ion-pair reverse-phase liquid chromatography under conditions that partially denature the DNA. Heteroduplexes, which are formed when different alleles combine and undergo melting characteristics distinct from homoduplexes that match perfectly, can be selectively separated and eluted from the latter [101].

8.2.1 SNP discovery by DHPLC

Scientists estimate that the human genome, comprising approximately 3 billion base pairs, harbors at least 1 million single-nucleotide polymorphisms, or SNPs. SNPs are genetic variations that occur commonly in the general population exceeding 1%. The significant interest in SNPs stems from their utility as markers for identifying disease-associated genes through linkage disequilibrium mapping. Identifying novel disease-linked genes not only aids in developing diagnostic tools but also presents potential targets for novel therapeutic agents.

Many laboratories currently use direct DNA sequencing for mutation detection and identification. However, due to its expense and time-consuming nature, alternative technologies have emerged that offer quicker and more cost-effective detection of genetic variations. These alternatives can be categorized into mutation scoring and mutation screening technologies. Mutation scoring, or genotyping, platforms target specific mutation sites or SNP locations to swiftly determine the sequence at those points.

Screening methods are employed to simultaneously identify both familiar and unfamiliar genetic variations. This article focuses on DHPLC, a platform technology effectively employed for mutation screening and SNP identification. DHPLC can also be adapted for genotyping and the analysis of gene expression, showcasing its versatility in genetic research and clinical applications [102].

Analysis of DHPLC conducted at a moderately denaturing temperature effectively isolates DNA fragments into homoduplexes and heteroduplexes when the injected sample contains multiple alleles. Typically, an experimental sample is combined with a reference sample of wild-type homozygous genotype prior to injection, facilitating the identification of sequence variations such as SNPs or pathological mutations [101].

8.2.2 DHPLC analysis of SNP genotyping

In the genotyping of SNPs from population specimens using DHPLC analysis, initial analysis involves testing all samples independently without combining them with a reference sample. This method enables the identification of both homozygous types and heterozygotes. A subsequent analysis involves combining each sample with a reference sample (either homozygous or heterozygous based on initial results), allowing precise genotyping of the targeted SNPs and simultaneous identification of any new SNPs present within the same PCR fragment [101].

8.3 DNA methylation analysis

A common trait found in numerous eukaryotic genomes involves the cytosine methylation at the carbon 5′ position within CpG dinucleotides. This methylation ordinarily occurs most frequently in CpG-rich regions known as "CpG islands," which are predominantly located in gene promoter regions or initial exons. The methylation of DNA plays a pivotal role in epigenetic regulation of gene expression and the preservation of genomic stability. As such, assessing the methylation condition of genomic DNA is essential for studying mechanisms of control of cell growth, differentiation specific to tissues, and the development of cancer.

Several methods are used to measure global DNA methylation, including HPLC, and its derivatives like LC-MS/MS, along with high-performance capillary electrophoresis and two-dimensional thin-layer chromatography. Many of these techniques assess a proxy for overall genomic methylcytosine, but there is a lack of certainty regarding their consistency and comparability [103].

8.3.1 DNA methylation analysis using HPLC

In HPLC, following DNA digestion, nucleotides are separated by size in chromatographic methods, allowing for the quantification of both cytosine and methylated cytosine. While this technique is highly quantitative and reproducible, it necessitates significant quantities of DNA and involves a demanding procedure for optimizing the assay. Consequently, alternative methods have been developed to assess overall methylation levels using smaller amounts of DNA and equipment that is more easily accessible. Various methods, such as quantitative methylation-specific PCR (MSP), Mass ARRAY, and pyrosequencing, have been employed for quantitative analysis of DNA methylation in research settings. However, these methods typically involve lengthy times of analysis, sophisticated procedures, and require costly, extensive equipment. Therefore, there is a critical requirement to develop an analytical system able to offer rapid, precise data, particularly for detecting DNA methylation abnormalities in diseased cells that need to be distinguished from contaminating cells.

An anion-exchange HPLC column has been developed to address these challenges specifically for detecting methylated DNA. Ion-exchange HPLC is commonly used to rapidly separate and analyze biopolymers like nucleic acids, proteins, and polysaccharides. A recent anion-exchange packing material has been developed to enhance the DNA methylation analysis by applying ion-exchange HPLC. This material combines electrostatic properties based on ion-exchange groups (quaternary ammonium and tertiary amino groups) with hydrophobic properties derived from hydrophobic monomers within the polymer particles.

As shown in **Table 1**, the different tests for assessing DNA methylation levels provide insights into their biological significance and implications for epigenetic research [104].

Assay	Biological significance	Quantity of DNA necessary (utilized in our assay)	Equipment needed
LINE1	~70,000 copies, corresponding to around 17% in the human genome	500 pg to 2 mg (250 ng)	Thermal cycler, pyrosequencer
Alu	~110,000 copies, corresponding to around 11% in the human genome	500 pg to 2 mg (250 ng)	Thermal cycler, pyrosequencer
LUMA	The percentage of CpG sites found within HpaII recognition sites (5′CCGG′3) in the human genome is 4.14% within transposable elements and 3.90% within unique sequences, totaling 8.04%	100-2500 ng (100 ng)	Incubator, pyrosequencer
HPLC-UV	Total 5mC in the human genome	1–5 mg (3 mg)	HPLC

Table 1.
Summary of different tests for evaluating overall DNA methylation levels, illustrating their biological significance [104].

8.3.2 DNA methylation analysis using DHPLC

Another HPLC technique known as denaturing HPLC (DHPLC) utilizes ion-pair RP-HPLC to identify DNA methylation under particular partial denaturation conditions. Several cancer researchers have documented its effectiveness in detecting DNA methylation in the promoter regions of tumor suppressor genes. Nevertheless, the method requires heteroduplex formation while analyzing, which prolongs the overall time of analysis. Moreover, minor variations in column temperature, even as small as 1°C, significantly alter peak shapes and retention times. This sensitivity can lead to imprecise measurement of DNA methylation levels, especially when methylated CpG positions vary, despite the total number of methylated CpGs being consistent.

In contrast, the anion-exchange HPLC method allows direct application of PCR products without time-consuming pre-analytical processes, completing the analysis within approximately 10 minutes. The recently developed column provides reproducible data using standard HPLC equipment, without requiring stringent temperature control [105].

8.3.3 DNA methylation analysis using LC-MS/MS

Chromatography has been employed for the separation of purines, pyrimidines, and the detection of changed deoxyribonucleosides. RP-HPLC techniques have been utilized for analysis of genome-wide methylation, albeit requiring microgram amounts of genomic DNA (5–50 μg), synthesis of 32P-labeled deoxyribonucleosides, and extended run times. In 1976, mass spectrometry techniques were introduced to enhance sensitivity in detecting and identifying 5-methylated cytosine residues in intact DNA without derivatization.

LC/MS has been recently recommended for assessing genomic methylation in DNA extracted from green algae, although it requires a substantial quantity of DNA (25 μg), 32P labeling of nucleosides, and an off-line mixture of HPLC and MS. This method involves HPLC separation and UV absorbance identification of deoxyribonucleosides, followed by ESI-MS analysis to individually analyze collected compounds. LC/MS allows for quantitative assessment of genomic DNA methylation status through complete hydrolysis of DNA and thorough elimination of residual RNA, enabling separation and detection of DNA bases and 5-methyl-2′-deoxycytidine by ESI-MS [103].

9. Conclusions

In conclusion, the integration of High-Performance Liquid Chromatography (HPLC) into clinical diagnostic procedures promises to significantly elevate the accuracy, sensitivity, and efficiency of disease detection and monitoring. By harnessing HPLC's robust capabilities for the precise separation, identification, and quantification of complex mixtures in biological samples, we can streamline the analysis of biomarkers, drugs, metabolites, and toxins in various bodily fluids. This not only facilitates rapid and reliable diagnosis of a wide range of diseases, including metabolic disorders and cancers but also enhances patient assessment and treatment decision-making processes. Therefore, the strategic incorporation of HPLC technology into clinical diagnostic protocols is essential for achieving improved patient outcomes through superior diagnostic precision and efficiency.

Acknowledgements

Special thanks to Pardis Clinical and Genetic Laboratory located in Mashhad, Iran, for their significant technical support and for providing the necessary facilities and all persons attending this research to carry out the study.

Conflict of interest

The authors declare no conflict of interest.

Author details

Fatemeh Keyfi[1,2*], Amin Alaei[1,2], Hadiseh Mirahmadi Daryasari[2,3], Ali Hakimi[4] and Pouya Gharavi[1]

1 Department of Medical Laboratory Science, Varastegan Institute for Medical Sciences, Mashhad, Iran

2 Division of Metabolic Disorder, Pardis Clinical and Genetic Laboratory, Mashhad, Iran

3 Department of Biochemistry, Faculty of Science, Ferdowsi University of Mashhad, Iran

4 Department of Clinical Biochemistry, Faculty of Medicine, Mashhad University of Medical Sciences, Mashhad, Iran

*Address all correspondence to: keifyf@varastegan.ac.ir

References

[1] Denoroy L, Zimmer L, Renaud B, Parrot S. Ultra high performance liquid chromatography as a tool for the discovery and the analysis of biomarkers of diseases: A review. Journal of Chromatography B. 2013;**927**:37-53

[2] Wajner M, Sitta A, Kayser A, Deon M, Groehs AC, Coelho DM, et al. Screening for organic acidurias and aminoacidopathies in high-risk Brazilian patients: Eleven-year experience of a reference center. Genetics and Molecular Biology. 2019;**42**(1 suppl 1):178-185

[3] Nikolin B, Imamović B, Medanhodzić-Vuk S, Sober M. High perfomance liquid chromatography in pharmaceutical analyses. Bosnian Journal of Basic Medical Sciences. 2004;**4**(2):5-9

[4] Shabir G. Validation of high-performance liquid chromatography methods for pharmaceutical analysis. Journal of Chromatography A. 2003;**987**:57-66

[5] Swartz M. HPLC detectors: A brief review. Journal of Liquid Chromatography and Related Technologies. 2010;**33**(9-12):1130-1150

[6] Forgács E, Cserháti T. Chromatography | principles. In: Caballero B, editor. Encyclopedia of Food Sciences and Nutrition. 2nd ed. Oxford: Academic Press; 2003. pp. 1259-1267

[7] Souza L, Alseekh S, Scossa F, Fernie A. Ultra-high-performance liquid chromatography high-resolution mass spectrometry variants for metabolomics research. Nature Methods. 2021;**18**(7):733-746

[8] Munkongdee T, Chen P, Winichagoon P, Fucharoen S, Paiboonsukwong K. Update in laboratory diagnosis of thalassemia. Frontiers in Molecular Biosciences. 2020;7:74

[9] Onyekere P, Egbuna C, Munir N, Daniyal M, Ahmed O, Găman M-A, et al. Vitamins and Minerals: Types, Sources and Their Functions. Philippines: Springer; 2020. pp. 149-172

[10] Papadoyannis IN, Tsioni GK, Samanidou VF. Simultaneous determination of nine water and fat soluble vitamins after SPE separation and RP-HPLC analysis in pharmaceutical preparations and biological fluids. Journal of Liquid Chromatography and Related Technologies. 1997;**20**(19):3203-3231

[11] Edwards KA, Tu-Maung N, Cheng K, Wang B, Baeumner AJ, Kraft CE. Thiamine assays-advances, challenges, and caveats. ChemistryOpen. 2017;**6**(2):178-191

[12] Li C, Liang C, Kong Z, Su Y, Ren W, Dong H, et al. Determination of vitamin K1, MK-4, MK-7, and D levels in human serum of postmenopausal osteoporosis women based on high stability LC-MS/MS: MK-7 may be a new marker of bone metabolism. Annals of Nutrition and Metabolism. 2023;**79**(3):334-342

[13] Yang Y, Lu D, Yang D, Yin S, Zhang J, Zheng B, et al. A rapid and sensitive HPLC-FLD method for determination of retinol and vitamin E isomers in human serum. Current Pharmaceutical Analysis. 2019;**15**(7):745-752

[14] Siluk D, Oliveira RV, Esther-Rodriguez-Rosas M, Ling S, Bos A, Ferrucci L, et al. A validated liquid chromatography method for the simultaneous determination of vitamins

A and E in human plasma. Journal of Pharmaceutical and Biomedical Analysis. 2007;**44**(4):1001-1007

[15] Deng P, Zhan Y, Chen X, Zhong D. Derivatization methods for quantitative bioanalysis by Lc–MS/MS. Bioanalysis. 2012;**4**(1):49-69

[16] Hymøller L, Jensen SK. Vitamin D analysis in plasma by high performance liquid chromatography (HPLC) with C(30) reversed phase column and UV detection—Easy and acetonitrile-free. Journal of Chromatography A. 2011;**1218**(14):1835-1841

[17] Górska RM. Chapter 5—Methods for assessment of vitamin K. In: Harrington D, editor. Laboratory Assessment of Vitamin Status. London, United Kingdom: Academic Press, Elsevier; 2019. pp. 107-147

[18] Zhang S-Q, Li J, Li L, Yuan X, Xu L, Shi Z-g. Fast separation of water-soluble vitamins by hydrophilic interaction liquid chromatography based on submicrometer flow-through silica microspheres. Food Chemistry. 2020;**307**:125531

[19] Saha DS, Saha B. Applications of Ion Exchange Resins in Vitamins Separation and Purification. Millersville, Pennsylvania: Materials Research Forum LLC; 2023. pp. 24-38

[20] Dabre R, Azad N, Schwämmle A, Lämmerhofer M, Lindner W. Simultaneous separation and analysis of water- and fat-soluble vitamins on multi-modal reversed-phase weak anion exchange material by HPLC-UV. Journal of Separation Science. 2011;**34**:761-772

[21] Midttun Ø, Ueland PM. Determination of vitamins A, D and E in a small volume of human plasma by a high-throughput method based on liquid chromatography/tandem mass spectrometry. Rapid Communications in Mass Spectrometry: RCM. 2011;**25**(14):1942-1948

[22] Basiri B, Sutton JM, Hanberry B, Zastre J, Bartlett M. Ion pair liquid chromatography method for the determination of thiamine (vitamin B1) homeostasis. Biomedical Chromatography: BMC. 2015;**30**(1):35-41

[23] Kurdi SE, Muaileq DA, Alhazmi HA, Bratty MA, El Deeb SJAP. Comparing monolithic and fused core HPLC columns for fast chromatographic analysis of fat-soluble vitamins. 2017;**67**(2):203-213

[24] Riphagen IJ, van der Molen JC, van Faassen M, Navis G, de Borst MH, Muskiet FA, et al. Measurement of plasma vitamin K1 (phylloquinone) and K2 (menaquinones-4 and -7) using HPLC-tandem mass spectrometry. Clinical Chemistry and Laboratory Medicine. 2016;**54**(7):1201-1210

[25] Crook MA. Chapter 6—Methods for assessment of thiamine (vitamin B1). In: Harrington D, editor. Laboratory Assessment of Vitamin Status. Cambridge, MA, USA: Academic Press; 2019. pp. 149-164

[26] Albahrani AA, Rotarou V, Roche PJ, Greaves RF. A simultaneous quantitative method for vitamins A, D and E in human serum using liquid chromatography-tandem mass spectrometry. The Journal of Steroid Biochemistry and Molecular Biology. 2016;**159**:41-53

[27] Zhang Y, Zhou W-E, Yan J-Q, Liu M, Zhou Y, Shen X, et al. A review of the extraction and determination methods of thirteen essential vitamins to the human body: An update from 2010. Molecules (Basel, Switzerland). 2018;**23**(6):1484

[28] Hiller-Sturmhöfel S, Bartke A. The endocrine system: An overview. Alcohol Health and Research World. 1998;**22**(3):153-164

[29] Kumari Y, Bai P, Waqar F, Asif AT, Irshad B, Raj S, et al. Advancements in the management of endocrine system disorders and arrhythmias: A comprehensive narrative review. Cureus. 2023;**15**(10):e46484

[30] Boschi S, De Iasio R, Mesini P, Bolelli GF, Sciajno R, Pasquali R, et al. Measurement of steroid hormones in plasma by isocratic high performance liquid chromatography coupled to radioimmunoassay. Clinica Chimica Acta; International Journal of Clinical Chemistry. 1994;**231**(1):107-113

[31] Bidlingmaier M, Wu Z, Strasburger C. Measurement of Peptide Hormones. 2008. p. 36-46.

[32] Samanidou V, Gika H, Papadoyannis I. Rapid HPLC analysis of thyroid gland hormones tri-iodothyronine (T-3) and thyroxine (T-4) in human biological fluids after SPE. Journal of Liquid Chromatography & Related Technologies. 2000;**23**:681-692

[33] Hjemdahl P. [63] Catecholamine measurements in plasma by high-performance liquid chromatography with electrochemical detection. Methods in Enzymology. 1987;**142**:521-534

[34] Acconcia F, Marino M. Steroid hormones: Synthesis, secretion, and transport. In: Belfiore A, LeRoith D, editors. Principles of Endocrinology and Hormone Action. Cham: Springer International Publishing; 2016. pp. 1-31

[35] Goodman HM. Chapter 1—Introduction. In: Goodman HM, editor. Basic Medical Endocrinology. 4th ed. San Diego: Academic Press; 2009. pp. 1-27

[36] Huang P, Li Z, Guo H. New advances in the regulation of leaf senescence by classical and peptide hormones. Frontiers in Plant Science. 2022;**13**:923136

[37] Kleine B, Rossmanith W. Hormones Derived by Amino Acid Conversion. Cham, Switzerland: Springer; 2016. pp. 237-245

[38] Zhang X, Xu H, Feng G, Yang L, Zhao L, Li R, et al. Sensitive HPLC-DMS/MS/MS method coupled with dispersive magnetic solid phase extraction followed by in situ derivatization for the simultaneous determination of multiplexing androgens and 17-hydroxyprogesterone in human serum and its application to patients with polycystic ovarian syndrome. Clinica Chimica Acta; International Journal of Clinical Chemistry. 2023;**538**:221-230

[39] Hameedat F, Hawamdeh S, Alnabulsi S, Zayed A. High performance liquid chromatography (HPLC) with fluorescence detection for quantification of steroids in clinical, pharmaceutical, and environmental samples: A review. Molecules (Basel, Switzerland). 2022;**27**(6):1807

[40] Rezk MR, Mohamed MF, Fathalla FA, Shehata MA. Selective determination of human growth hormone (somatropin) in the presence of its chemical degradation products. Journal of AOAC International. 2017;**100**(4):1023-1028

[41] Gabler J, Wang S. Quantification of metanephrine and normetanephrine in urine using liquid chromatography-tandem mass spectrometry. Methods in Molecular Biology (Clifton, NJ). 2016;**1378**:149-157

[42] Capp MW, Simonian MH. Separation of the major adrenal steroids by reversed-phase high-performance

liquid chromatography. Analytical Biochemistry. 1985;**147**(2):374-381

[43] Lin L, Meltzer N, Honigberg I. Normal phase HPLC with EC detection: 1,25-dihydroxy vitamin D3 and 25-hydroxy-16-ene-23-yne vitamin D3. Journal of Liquid Chromatography. 2006;**16**:3093-3100

[44] Camperi J, Combes A, Fournier T, Pichon V, Delaunay N. Analysis of the human chorionic gonadotropin protein at the intact level by HILIC-MS and comparison with RPLC-MS. Analytical and Bioanalytical Chemistry. 2020;**412**(1):4423-4432

[45] Gu Q, Shi X, Yin P, Gao P, Wu N, Xu G. Analysis of catecholamines and their metabolites in adrenal gland by liquid chromatography tandem mass spectrometry. Analytica Chimica Acta. 2008;**609**:192-200

[46] Mant CT, Chen Y, Yan Z, Popa TV, Kovacs JM, Mills JB, et al. HPLC analysis and purification of peptides. Methods in Molecular Biology (Clifton, NJ). 2007;**386**:3-55

[47] Garrod AE. Inborn Errors of Metabolism. 2nd ed. London: Henry Frowde and Hodder & Stoughton; 1923

[48] Hazan G, Hershkovitz E, Staretz-Chacham O. Incidence of inherited metabolic disorders in southern Israel: A comparison between consanguinity and non-consanguinity communities. Orphanet Journal of Rare Diseases. 2020;**15**(1):331

[49] Sharma G, Attri SV, Behra B, Bhisikar S, Kumar P, Tageja M, et al. Analysis of 26 amino acids in human plasma by HPLC using AQC as derivatizing agent and its application in metabolic laboratory. Amino Acids. 2014;**46**(5):1253-1263

[50] Yang X, Ma Y, Li N, Cai H, Bartlett M. Development of a novel method for the determination of acyl-CoA compounds by liquid chromatography mass spectrometry to probe the metabolism of fatty acids. Analytical Chemistry. 2016;**89**(1):813-821

[51] Want EJ, Wilson ID, Gika H, Theodoridis G, Plumb RS, Shockcor J, et al. Global metabolic profiling procedures for urine using UPLC-MS. Nature Protocols. 2010;**5**(6):1005-1018

[52] Churchwell MI, Twaddle NC, Meeker LR, Doerge DR. Improving LC-MS sensitivity through increases in chromatographic performance: Comparisons of UPLC-ES/MS/MS to HPLC-ES/MS/MS. Journal of Chromatography. B, Analytical Technologies in the Biomedical and Life Sciences. 2005;**825**(2):134-143

[53] Ramsay J, Morton J, Norris M, Kanungo S. Organic acid disorders. Annals of Translational Medicine. 2018;**6**(24):472

[54] Haas D, Hauke J, Schwarz KV, Consalvi L, Trefz FK, Blau N, et al. Differences of phenylalanine concentrations in dried blood spots and in plasma: Erythrocytes as a neglected component for this observation. Metabolites. 2021;**11**(10):680

[55] Williams RA, Mamotte CD, Burnett JR. Phenylketonuria: An inborn error of phenylalanine metabolism. The Clinical Biochemist Reviews. 2008;**29**(1):31-41

[56] Regier DS, Greene CL. Phenylalanine Hydroxylase Deficiency. In: Adam MP, Feldman J, Mirzaa GM, et al., editors. GeneReviews®. Seattle (WA): University of Washington, Seattle; 2000 [Updated

2017 Jan 5]. Available from: https://www.ncbi.nlm.nih.gov/books/NBK1504/

[57] van Wegberg AMJ, MacDonald A, Ahring K, Bélanger-Quintana A, Blau N, Bosch AM, et al. The complete European guidelines on phenylketonuria: Diagnosis and treatment. Orphanet Journal of Rare Diseases. 2017;**12**(1):162

[58] Mo XM, Li Y, Tang AG, Ren YP. Simultaneous determination of phenylalanine and tyrosine in peripheral capillary blood by HPLC with ultraviolet detection. Clinical Biochemistry. 2013;**46**(12):1074-1078

[59] Kand'ár R, Záková P. Determination of phenylalanine and tyrosine in plasma and dried blood samples using HPLC with fluorescence detection. Journal of Chromatography. B, Analytical Technologies in the Biomedical and Life Sciences. 2009;**877**(30):3926-3929

[60] Shi Z, Li H, Li Z, Hu J, Zhang H. Pre-column derivatization RP-HPLC determination of amino acids in Asparagi Radix before and after heating process. IERI Procedia. 2013;**5**:351-356

[61] Babu SV, Shareef MM, Shetty AP, Shetty KT. HPLC method for amino acids profile in biological fluids and inborn metabolic disorders of aminoacidopathies. Indian Journal of Clinical Biochemistry: IJCB. 2002;**17**(2):7-26

[62] Schwarz EL, Roberts WL, Pasquali M. Analysis of plasma amino acids by HPLC with photodiode array and fluorescence detection. Clinica Chimica Acta; International Journal of Clinical Chemistry. 2005;**354**(1-2):83-90

[63] Sharer JD, De Biase I, Matern D, Young S, Bennett MJ, Tolun AA. Laboratory analysis of amino acids, 2018 revision: A technical standard of the American College of Medical Genetics and Genomics (ACMG). Genetics in Medicine: Official Journal of the American College of Medical Genetics. 2018;**20**(12):1499-1507

[64] Fuenzalida K, Valiente A, Faúndez A, Guerrero P, Soto V, Leal-Witt MJ, et al., editors. Quantitative determination of branched-chain amino acids in dried blood spot samples by LC-MSMS and its application in diagnosis and follow-up of chilean patients with maple syrup urine disease. Journal of Inborn Errors of Metabolism and Screening, Brazil. 2021

[65] Shennar HK, Al-Asmar D, Kaddoura A, Al-Fahoum S. Diagnosis and clinical features of organic acidemias: A hospital-based study in a single center in Damascus, Syria. Qatar Medical Journal. 2015;**2015**(1):9

[66] Ghoraba D, Mohamed M, Zaki O. Screening of diseases associated with abnormal metabolites for evaluation of HPLC in organic aciduria profiling. Egyptian Journal of Medical Human Genetics. 2013;**15**(1):69-78

[67] Cunha SC, Fernandes JO, Ferreira IM. HPLC/UV determination of organic acids in fruit juices and nectars. European Food Research and Technology. 2002;**214**(1):67-71

[68] Colin H, Guiochon G. Introduction to reversed-phase high-performance liquid chromatography. Journal of Chromatography A. 1977;**141**(3):289-312

[69] Mandrioli R, Albani F, Casamenti G, Sabbioni C, Raggia M. Simultaneous high-performance liquid chromatography determination of carbamazepine and five of its metabolites in plasma of epileptic patients. Journal of Chromatography. B, Biomedical Sciences and Applications. 2001;**762**:109-116

[70] Sander LC, Wise SA. Synthesis and characterization of polymeric C18 stationary phases for liquid chromatography. Analytical Chemistry. 1984;**56**(3):504-510

[71] Deidda M, Piras C, Bassareo P, Cadeddu C, Mercuro G. Metabolomics, a promising approach to translational research in cardiology. IJC Metabolic and Endocrine. 2015;**9**(1):31-38

[72] Lee MY, Hu T. Computational methods for the discovery of metabolic markers of complex traits. Metabolites. 2019;**9**(4):66

[73] Zhao YY, Cheng XL, Vaziri ND, Liu S, Lin RC. UPLC-based metabonomic applications for discovering biomarkers of diseases in clinical chemistry. Clinical Biochemistry. 2014;**47**(15):16-26

[74] Zhao X, An X, Yang C, Sun W, Ji H, Lian F. The crucial role and mechanism of insulin resistance in metabolic disease. Frontiers in Endocrinology (Lausanne). 2023;**14**(1):1149239

[75] Wu T, Qiao S, Shi C, Wang S, Ji G. Metabolomics window into diabetic complications. Journal of Diabetes Investigation. 2018;**9**(2):244-255

[76] Galicia-Garcia U, Benito-Vicente A, Jebari S, Larrea-Sebal A, Siddiqi H, Uribe KB, et al. Pathophysiology of type 2 diabetes mellitus. International Journal of Molecular Sciences. 2020;**21**(17):6275

[77] Hameed A, Mojsak P, Buczynska A, Suleria HAR, Kretowski A, Ciborowski M. Altered metabolome of lipids and amino acids species: A source of early signature biomarkers of T2DM. Journal of Clinical Medicine. 2020;**9**(7):2257

[78] Kohne E. Hemoglobinopathies: Clinical manifestations, diagnosis, and treatment. Deutsches Ärzteblatt International. 2011;**108**(31-32):532-540

[79] Akhavan-Niaki H, Youssefi Kamangari R, Banihashemi A, Kholghi Oskooei V, Azizi M, Tamaddoni A, et al. Hematologic features of alpha thalassemia carriers. International Journal of Molecular and Cellular Medicine. 2012;**1**(3):162-167

[80] Fucharoen S, Viprakasit V. Hb H disease: Clinical course and disease modifiers. Hematology. American Society of Hematology. Education Program. 2009;**2009**(1):26-34

[81] Origa R. β-Thalassemia. Genetics in Medicine: Official Journal of the American College of Medical Genetics. 2017;**19**(6):609-619

[82] Sharma P, Das R. Cation-exchange high-performance liquid chromatography for variant hemoglobins and HbF/A2: What must hematopathologists know about methodology? World Journal of Methodology. 2016;**6**(1):20-24

[83] Warghade S, Britto J, Haryan R, Dalvi T, Bendre R, Chheda P, et al. Prevalence of hemoglobin variants and hemoglobinopathies using cation-exchange high-performance liquid chromatography in central reference laboratory of India: A report of 65779 cases. Journal of Laboratory Physicians. 2018;**10**(1):73-79

[84] Khera R, Singh T, Khuana N, Gupta N, Dubey AP. HPLC in characterization of hemoglobin profile in thalassemia syndromes and hemoglobinopathies: A clinicohematological correlation. Indian Journal of Hematology & Blood Transfusion: An Official Journal of Indian Society of Hematology and Blood Transfusion. 2015;**31**(1):110-115

[85] Di Pierro E, De Canio M, Mercadante R, Savino M, Granata F, Tavazzi D, et al. Laboratory diagnosis of porphyria. Diagnostics (Basel, Switzerland). 2021;**11**(8):1343

[86] Macours P, Cotton F. Improvement in HPLC separation of porphyrin isomers and application to biochemical diagnosis of porphyrias. Clinical Chemistry and Laboratory Medicine. 2006;**44**(12):1433-1440

[87] Lim CK, Rideout JM, Wright DJ. Separation of porphyrin isomers by high-performance liquid chromatography. The Biochemical Journal. 1983;**211**(2):435-438

[88] Aarsand AK, Petersen PH, Sandberg S. Estimation and application of biological variation of urinary delta-aminolevulinic acid and porphobilinogen in healthy individuals and in patients with acute intermittent porphyria. Clinical Chemistry. 2006;**52**(4):650-656

[89] Rifai N. Tietz Textbook of Clinical Chemistry and Molecular Diagnostics—E-Book. Saunders; Elsevier, Health Science; 2017

[90] Sankar R, Rajyalakshmi G, Chapala D, Rao G. Instant tips for right and effective approach to solve HPLC trouble shooting. Journal of Chemical and Pharmaceutical Sciences. 2014;**7**:259-274

[91] Ansari A, Gupta R, Jhanwar M, Kaushik P, Sharma D, Shahid M, et al. HPLC & It's utilization in disease diagnosis. Asian Journal of Pharmaceutical Research and Development. 2020;**8**:114-116

[92] Freudenberger K, Hilbig U, Gauglitz G. Recent advances in therapeutic drug monitoring of immunosuppressive drugs. TrAC Trends in Analytical Chemistry. 2016;**79**(1):257-268

[93] Schellenberg F, Wielders J, Anton R, Bianchi V, Deenmamode J, Weykamp C, et al. IFCC approved HPLC reference measurement procedure for the alcohol consumption biomarker carbohydrate-deficient transferrin (CDT): Its validation and use. Clinica Chimica Acta; International Journal of Clinical Chemistry. 2017;**465**:91-100

[94] Venkatakrishna KL, Kartha VB, Pai KM, Krishna CM, Ravikiran O, Kurian J, et al. HPLC-LIF for early detection of oral cancer. Current Science. 2003;**84**:551-557

[95] Bhat S, Rai L, Krishnanand B, Mahato K, Kartha V, Chidangil S. Protein profile study of the cervical cancer using HPLC-LIF—Art. no. 60920V. In: Proceedings of SPIE—The International Society for Optical Engineering; United States. 2006. p. 6092

[96] Fan X, Bai J, Shen P. Diagnosis of breast cancer using HPLC metabonomics fingerprints coupled with computational methods. In: Conference Proceedings: Annual International Conference of the IEEE Engineering in Medicine and Biology Society Annual Conference; Denmark. 2005. pp. 6081-6084

[97] Menon S, Bhat S, Kumar K, Kurien J, Krishnanand B, Mahato K, et al. Protein profile study of breast cancer tissues using HPLC-LIF—A pilot study—Art. no. 64300W. In: Proceedings of SPIE—The International Society for Optical Engineering; United States. 2007. p. 6430

[98] Zusman I, Zusman R, Sandler B, Korol D, Gurevich P, Smirnoff P, et al. HPLC determination of serum levels of soluble p53 antigen as a new method for colon cancer detection, and its clinical

implication. International Journal of Oncology. 1996;**8**(6):1249-1256

[99] Yang J, Xu G, Zheng Y, Kong H, Pang T, Lv S, et al. Diagnosis of liver cancer using HPLC-based metabonomics avoiding false-positive result from hepatitis and hepatocirrhosis diseases. Journal of Chromatography. B, Analytical Technologies in the Biomedical and Life Sciences. 2004;**813**(1-2):59-65

[100] Issaq HJ, Nativ O, Waybright T, Luke B, Veenstra TD, Issaq EJ, et al. Detection of bladder cancer in human urine by metabolomic profiling using high performance liquid chromatography/mass spectrometry. The Journal of Urology. 2008;**179**(6):2422-2426

[101] Han W, Yip S, Yap M. Using denaturing HPLC for SNP discovery and genotyping, and establishing the linkage disequilibrium pattern for the all-trans-retinol dehydrogenase (RDH8) gene. Journal of Human Genetics. 2004;**49**:16-23

[102] Frueh FW, Noyer-Weidner M. The use of denaturing high-performance liquid chromatography (DHPLC) for the analysis of genetic variations: Impact for diagnostics and pharmacogenetics. Chemistry and Laboratory Medicine. 2003;**41**(4):452-461

[103] Friso S, Choi SW, Dolnikowski GG, Selhub J. A method to assess genomic DNA methylation using high-performance liquid chromatography/electrospray ionization mass spectrometry. Analytical Chemistry. 2002;**74**(17):4526-4531

[104] Lisanti S, Omar WA, Tomaszewski B, De Prins S, Jacobs G, Koppen G, et al. Comparison of methods for quantification of global DNA methylation in human cells and tissues. PLoS One. 2013;**8**(11):e79044

[105] Yotani T, Yamada Y, Arai E, Tian Y, Gotoh M, Komiyama M, et al. Novel method for DNA methylation analysis using high-performance liquid chromatography and its clinical application. Cancer Science. 2018;**109**(5):1690-1700

Chapter 6

High-Throughput Chromatography for Clinical Proteomics Applications

Goran Mitulović and Jadranka Miletić Vukajlović

Abstract

It is becoming increasingly important to use high-throughput analysis of clinical samples to monitor the progression or regression of disease or the concentration of therapeutic agents during personalized treatment. The use of micro and capillary chromatography is well established, but nano-LC can provide significantly higher sensitivity, especially for analyzing therapeutic monoclonal antibodies. Although the nano-LC has been considered a slow procedure, new separation columns, for example, monolithic columns, can provide both high sample throughput and high sensitivity. When hyphenated with a high-end mass spectrometer, the combination can be used for screening and targeted analyses of clinical samples. Furthermore, a combination of different, orthogonal separation methods can be used to increase the selectivity and sensitivity of the analysis. Some actual research and applied topics for clinical sample analyses will be discussed.

Keywords: high-throughput, proteomics, LC-MS, precision medicine, biomarker discovery

1. Introduction

Proteomics has experienced rapid growth in the past two decades. Remarkably, the development of proteomics methods for therapeutic applications has surged. Proteomics comprehensively studies protein interactions and functions, the structure of the amino acid chains and peptide backbones, and, finally, the role of proteins in cellular activities and how the changes in the proteome affect biological activity [1].

Proteomics provides supplementary insights to genomics and transcriptomics and plays a vital role in identifying and constructing a comprehensive map of interlinked pathways, networks, and molecular systems [2]. The great advantage of proteomics is that it offers a more profound understanding of an organism's structure and biological functions than genomics does. However, it presents heightened complexity due to the dynamic nature of protein expression influenced by temporal and environmental factors. With an estimated abundance nearing one million human proteins, many of which undergo post-translational modifications (PTMs), including phosphorylation and protein degradation, proteomics reveals a broader spectrum of proteins compared to the approximately 26,000–31,000 proteins encoded by the human genome [3].

IntechOpen

Thus, the data obtained only from the genome cannot fully picture the disease mechanisms with a comprehensive understanding [4].

Proteins, functioning as the primary units driving cellular activities, exhibit dynamic changes in abundance and modifications during processes like cell differentiation, environmental shifts, aging, disease progression, and responses to drug treatments. Therefore, analyses of proteins in cells, body fluids, or tissues are crucial for advancing basic research, medicine, and biotechnology [5, 6]. Proteome analyses aim to identify changes in protein expression and explore post-translational modifications, protein-protein interactions, cellular and sub-cellular distribution, and temporal expression patterns.

The efficacy of proteomics for identifying biological pathways and disease mechanisms is now widely established. Proteomics has unveiled numerous potential drug targets for diverse diseases [7]. There are a variety of proteomics techniques from western blot, as well as multidimensional protein identification technology [8], stable isotope labeling with amino acids in cell culture [9], immunohistochemistry (IHC) staining, enzyme-linked immunosorbent assay (ELISA) to high-throughput methods such as tissue microarray (TMA), protein pathway array and mass spectrometry (MS) [10].

Regardless of the technique employed, these comprehensive proteomic methodologies can be categorized into three main phases: discovery, network-analysis, and clinical proteomics [4] (**Figure 1**).

The discovery phase marks the beginning of each, aiming to identify amino acid sequences and elucidate unknown protein structures with qualification [11]. Methods that have been widely used to identify and quantify new proteins in the initial discovery stage are mass spectrometry (MS)-based methods, single-molecule proteomics (SMP), and single-cell proteomics (SCP). Subsequently, during the network analysis phase, global signaling networks were constructed, and the relationships among established proteins were to uncover potential biomarkers, which are then subjected to verification. The protein pathway array (PPA) is a high-throughput method designed to investigate the regulation of protein-protein interactions, pathway-pathway interactions, and biological functions. It aims to determine the placement of newly discovered proteins within cell signaling networks. Finally, in the clinical proteomics phase [12], several effective high-throughput techniques exist for clinical validation after the proteomic markers are validated using tissue microarray (TMA).

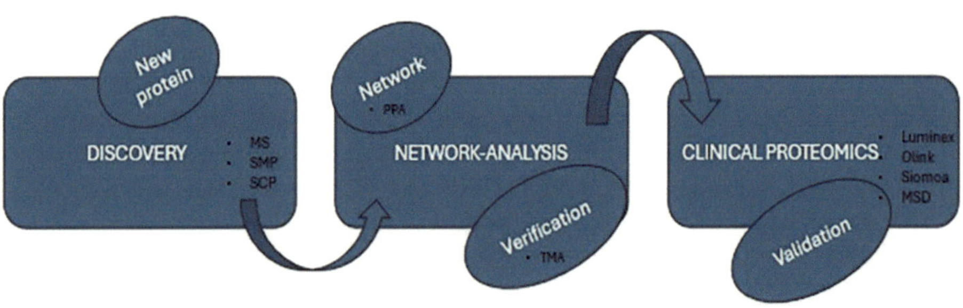

Figure 1.
Global proteomics process. In the initial discovery stage, mass spectrometry (MS)-based methods, such as single-molecule proteomics (SMP) and single-cell proteomics (SCP), have been broadly used to identify and quantify unknown proteins. The protein pathway array (PPA) is a high-throughput method used to investigate the regulation of protein-protein interactions, pathway-pathway interactions, and biological functions. It aids in determining the placement of newly identified proteins within cell signaling networks. Luminex, Olink, Simoa and Meso-scale Discovery (MSD) are efficient high-throughput techniques for clinical validation following the verification of proteomic markers using tissue microarray (TMA).

MS has emerged as the primary technique for conducting high-throughput proteome analysis across various biological systems. Due to the complex protein composition and wide range of protein abundance in plasma [13, 14], it is common practice to enhance identification depth by eliminating highly abundant proteins and fractionating either proteins or peptides [15]. However, when it comes to large-scale clinical samples, the elaborate sample preparation process not only consumes considerable time but also can introduce technical biases [16] and influence the analysis precision, especially for quantitative approaches. Therefore, automating and integrating the sample preparation steps in plasma proteomics is an attractive strategy to enhance the efficiency and reproducibility of clinical assays, as it facilitates easy standardization and scalability. Large-scale proteomics experiments often necessitate a trade-off between achieving depth in proteomic analysis and maintaining precision and throughput. While all three factors hold significance, their importance varies depending on the biological inquiry at hand. For instance, achieving depth becomes crucial in proteomic investigations if the objective is to detect specific low-abundance proteins (such as certain transcription factors or extracellular signaling molecules like cytokines). However, the significance of depth diminishes with increasing sample size as biological systems operate within intricate networks [17, 18]. Consequently, biological responses often entail concentration changes across numerous proteins [19–21], not all of which must be quantified to discern the response. Integrating multiparametric statistics and machine learning into biomedicine, alongside a deeper understanding of biological networks, instills a conceptual shift. Recognizing the organization of biological systems within networks reduces the imperative to quantify every individual protein each time. Instead, there is a growing emphasis on investigating diverse conditions and deciphering "signatures" that encapsulate aspects of the response [22–24]. In preparing high-throughput proteomics experiments, it is essential first to establish a study design to avoid any correlation between biological signals and/or sample groups with technical factors, as such correlations could result in misinterpretations of the data [25–27]. In clinical research, it is essential to account for additional variables that lie beyond the control of the analytical laboratory. The effectiveness of large-scale biomarker investigations relies on obtaining well-defined and equitable sample cohorts, which can be challenging regarding accessibility and recruitment. Moreover, ensuring standardized sample collection techniques within a study is crucial to prevent potential confounding with disease/control groups [14, 28, 29]. Sample preparation is critical in any proteomics experiment and typically involves multistep procedures. Even with advanced instrumentation and data analysis techniques, inadequate sample quality profoundly impacts all subsequent processes.

Chromatography plays a crucial role in quantitative LC-MS workflows, significantly influencing the efficiency and consistency of proteomics experiments. Factors such as gradient length, column wash, and equilibration times dictate the overall duration of the process. Chromatographic systems aim for minimal overhead times to accommodate shorter LC gradients, including rapid column wash and equilibration. This goal has been realized by transitioning from the typical nanoliter flow rates to capillary or microflow rates (<50 μl/min) [30–33] or even analytical flow rates (800 μl/min) [34–36]. Higher flow rates decrease dead times and solvent delay, leading to shorter wash and equilibration times while also enhancing robustness and spray stability and reducing carry-over [31, 32, 34, 37]. However, the drawback of high-flow chromatography is increased sample dilution, necessitating larger sample volumes per injection compared to lower flow rates. In general, the implementation of short gradient chromatography is fueled by advancements in chromatographic systems and novel column technologies, which enable high peak capacities even during brief separation periods. Modern UHPLC systems and columns with increased pressure limits allow for the utilization of small particles, promoting fast and efficient separations. Moreover, innovations such as micropillar array columns (μPAC) [38] or core-shell particles [39] further enhance the efficiency of peptide separations.

MS has emerged as a fundamental and widely utilized tool for identifying proteins and their isoforms and quantifying post-translational modifications. This is achieved through direct analysis of fragments or by detecting the specific proteolytic activity responsible for their generation. The primary impact of MS lies in its ability to identify and detect intact proteins or specific sets of composite or surrogate peptides, which poses a considerable challenge or impossibility for traditional immunoassays in quantitative proteomics. MS can be integrated with various separation and prefractionation methods to enhance protein/peptide identification accuracy and yield. While two-dimensional polyacrylamide gel electrophoresis (2D-PAGE) has historically served as a conventional method in proteomics research, gel-based techniques are often laborious and time-consuming, rendering them unsuitable for high-throughput proteomics applications. Conversely, LC, particularly high-performance liquid chromatography (HPLC), offers continuous separation of numerous proteins from complex mixtures and can be integrated with MS as LC-MS to enhance throughput [40, 41]. Reverse-phase liquid chromatography (RPLC) is the most frequently utilized method among these chromatographic methods. There are two complementary methods for peptide measurement in MS. Targeted MS employs stable-isotope standard peptides as reference markers to achieve absolute quantification of the peptides present in the sample. In contrast, untargeted MS relies on the intensity of peptide ions as a semi-quantitative indicator of peptide abundance. In the case of plasma, where peptides from low-abundance proteins may be overshadowed by those from highly abundant proteins, achieving accurate measurement across a broader spectrum of protein abundance concentrations typically necessitates extensive sample prefractionation. This limitation restricts MS-based analyses to smaller-scale studies [42]. Untargeted MS commonly utilizes data-dependent acquisition (DDA) MS methods, which prioritize the analysis of the most prominent peptide ions and have been extensively employed. With DDA techniques selecting individual precursor ions for MS/MS fragmentation, the MS/MS scan speed restricts the proteomic depth achievable, particularly during rapid gradients. Despite the ability of the latest generation time-of-flight (TOF) instruments to perform over 100 MS/MS scans per second [43], this limitation significantly hinders the effectiveness of DDA when coupled with fast chromatography. Advancements in MS instrumentation have accelerated scanning speeds and heightened sensitivity. These progressions have given rise to a next-generation proteomic approach termed Data-Independent Acquisition (DIA)-MS [44–49] (**Figure 2**) or SWATH-MS, which offers superior reproducibility and sensitivity compared to conventional DDA-MS [48]. In contrast to DDA-MS, DIA-MS involves fragmenting all precursor ions identified in an MS1 survey scan, with fragment ions accumulated within a set number of broad isolation windows spanning the entire mass-to-charge ratio (m/z) range and widening the dynamic range. Although DIA requires less intricate pre-analytical sample handling compared to DDA, it entails increased time for data collection and subsequent bioinformatic analyses to align the expanded list of targets with comprehensive protein libraries. HT DIA methodologies are advanced by developments in MS instruments that enhance ion transmission efficiencies and shorten duty cycles. The width of Q1 isolation is contingent upon the MS/MS duty cycle in DIA. Shorter duty cycles

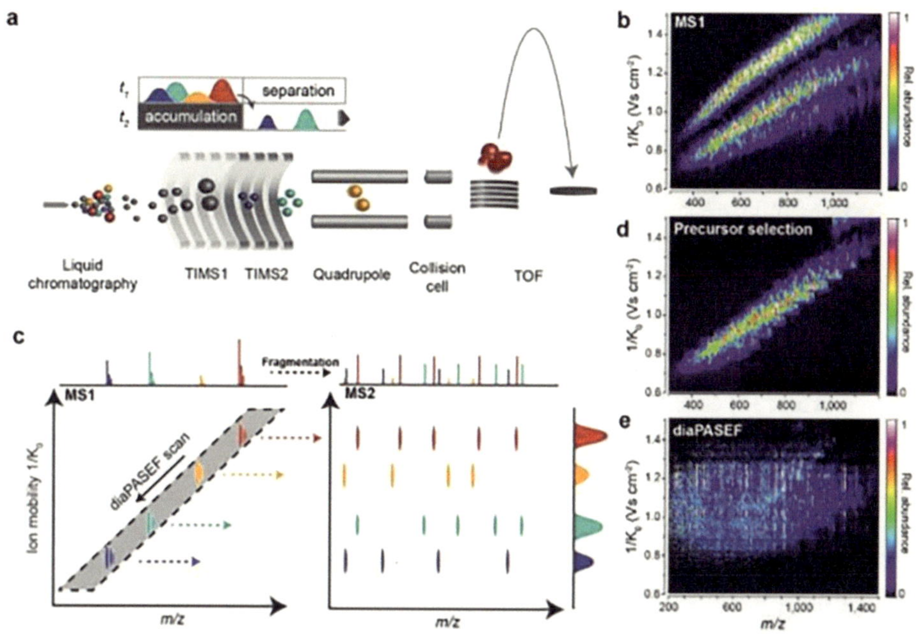

Figure 2.
Data-Independent Acquisition (DIA) is used for deep proteomics analysis of biological samples. Source: Meier et al. [48].

facilitate more narrow isolation windows, thereby boosting selectivity—the capacity to differentiate signals from distinct peptides.

The DIA approach provides significant improvements in proteomic depth, data completeness, robustness, and quantification accuracy, especially in experiments utilizing fast chromatographic gradients and optimized and advanced MS methods executed on the significantly improved hardware [50–52].

MS-based single-cell proteomics has been particularly successful for identifying cell heterogeneities in tissues, especially in solid tumors. The single-cell proteomics analysis enables identification of different cell development stages and understanding the biological processes undergoing in cells during the tumor growth or during the therapy. Through advanced automated sample preparation methods and the utilization of label-free or multiplexed data collection on highly sensitive instruments, researchers can now routinely identify and quantify between 1000 and 1500 proteins per cell. However, these numbers still fall short compared to the total count of unique proteins and proteoforms within a cell. The present status of MS-based single-cell proteomics resembles the initial phases of next-generation sequencing.

2. High-throughput LC for clinical analyses

Nano-liquid chromatography (nano-LC) combined with mass spectrometry is the primary tool for separating and identifying peptides generated mainly through enzymatic digestion. It provides excellent sensitivity and reproducibility even for the lowest sample amounts, for example, for the analysis of single cells [47, 53–59]. Spatial proteomics, a powerful approach that simultaneously obtains spatial information on thousands of proteins from biological samples, requires hundreds of micro-sections and weeks or even months to analyze several hundreds of tissue sections. All that makes the current approaches and analytical methods time-consuming, complex, and impractical.

There are several studies being conducted to develop high-throughput methods and technologies accessible to clinical proteomics research. Some of them have shown that the microflow LC-MS/MS systems operated at flow rates ranging from 50 to 800 μL/min and using separation columns with 2.1 mm inner diameter are robust and provide reproducible results [60].

The COVID-19 pandemic provoked the development of many analytical technologies and methods to address the identification of the virus, investigate the mechanism of action during the infection, and, ultimately, analyze many patient samples. Given the high number of infections, a high-throughput method was urgently needed. Messner et al. [34] established an analytical method for separating and identifying peptides from neat human plasma. By combining a 2.1 mm inner-diameter column of only 5 cm length, it is, in theory, possible to analyze up to 384 samples a day, including the time for sample preparation. The authors report detecting and quantifying at least 44 protein biomarkers approved by the FDA. Furthermore, although the number of identified proteins is essential for large-scale experiments and the targeted analysis and identification of either peptides or proteins, consistent quantification values for peptides and proteins are of significantly higher importance. The authors report that the detection of 311 protein groups and uniquely identifying proteins was measured with a remarkably high data completeness of 87% and with at least five identified peptides per protein.

Another report describes using a 30-second gradient generated and applied at 800 μL/min. This high-speed separation identified more than 1900 protein groups from 5 μg of tryptically digested K562 proteins [61]. Szyrwiel et al. reported the detection of more than 5000 proteins from 3 μg of trypsinized K562 using a 3—or 5-minute gradient at a flow rate of 500 μL/min [62].

The single-cell proteomics analysis has made remarkable progress in the past few years. At least several hundred cells must still be analyzed to discover the heterogeneity in cell types. For single-cell analysis, the sensitivity of the nanoflow LC-MS/MS is a priority due to the low protein amount in the single cell. Although the protein amount in a cell depends on its size, it is usually assumed that 250 pg. of proteins are present in a single mammalian cell. Of course, the amount of protein in a cell might and will differ for cells of different sizes or maturity stages.

The highest identification rates can be achieved using nano-LC with low or ultra-low flow rates and separation columns with small inner diameters. However, achieving high throughput with a nano-LC and low flow rates is impossible with the current setup. Significant development focuses on enhancing the sample throughput of LC-MS/MS analysis.

In the clinical approach, proteomics can be used for explorative and targeted studies. While explorative studies do not necessarily need the high-throughput approach, targeted analyses are usually employed to detect and quantify putative biomarkers in large sample cohorts. For the targeted approaches, for example, selected reaction monitoring (SRM) or parallel reaction monitoring (PRM), separation columns with large ID, for example, 500 μm or 1 mm and 2.1 mm, are increasingly used [63, 64].

A data-independent acquisition (DIA) [48] approach aims to catalog the peptides in a biological sample systematically. Capillary columns with an inner diameter of 300 μm ID are used. These columns seem to compromise robustness and sensitivity, which are needed to analyze biological samples, especially clinical samples [30, 65].

The sensitivity of mass spectrometers has dramatically improved in recent years due to, for example, increased ionization efficiency, improved and enhanced ion transfer, and advances in DDA (data-dependent analysis) due to the introduction of new fragmentation approaches such as PASEF (parallel serial fragmentation) combined with ion mobility separation and fast scanning TOF (time-of-flight) detector [44, 56, 66–71].

However, despite mass spectrometers' increased sensitivity and analysis speed, LCs separation speed limits sample throughput. Columns with larger inner diameters operated at higher flow rates were employed to overcome that limitation.

An example of a significant advance in clinical proteomics is the results published by Lenčo et al. [72], who reported the identification of about 2800 human proteins in 60 min from 2 μg of HeLa protein digests using an online LC-MS/MS method employing a 1 mm ID column. Compared to that, a recent multicentric experiment used nanoflow, a fast five-minute gradient on a 5 cm separation column, and 200 ng tryptically digested K562. The authors identified, on average, 7072 protein groups and 99,835 peptides at 1% FDR using library-free data processing [73]. The results show that even the nanoflow approach can be used for high sample throughput, which can be applied to clinical proteomics. For clinical analysis, it is also essential to have low variations in quantitative results, which was given in the current experiment with 12.1% for 110 runs.

Of course, running and analyzing standardized samples such as HeLa or K562 and a completely different approach to analyzing patient samples is significantly more straightforward.

Analysis of clinical samples implies using biological material obtained from patients who often suffer from multiple conditions and are significantly diverse. That is when trap columns are increasingly used for sample loading and clean-up. Trap columns save significant time, which is lost during direct sample injection. They enable optimal use of a short liquid chromatography gradient paired with rapid mass spectrometry data acquisition, which can quantify a moderate set of analytes. High-throughput proteomics analyses reproducibly can be used for sample profiling at a limited depth. Kreimer et al. [74] showed that trap columns can significantly boost the performance and the sample throughput. The authors used the dual-trap and single-column setup to maximize the throughput of blood and cell lysates. The total analysis time was 15 min per sample when the analysis path was parallelized using trap column cleaning and conditioning with sample loading and desalting on one trap column. The second trap column was switched in line with the separation column for sample analysis. The authors report achieving 90% utilization of the instrument time and applying a 9.5 μL/min microflow to analyze dried blood spots. The approach reproducibly quantified 300–400 proteins and over 6000 precursor ions.

Using the same approach, the authors reported the identification of over 4000 proteins when analyzing cell lysates. A Bruker TIMS-TOF Pro mass spectrometer with a DIA approach was used in both cases.

The same author used the dual-trap column setup to analyze heterogeneous cell populations [75]. In clinics, rare cell types are needed to be characterized for diagnostic purposes. In heterogeneous cell populations, rare cell types must be extracted from tissue or organs, and profiling hundreds of individual cells is needed. The previous study shows that parallelized nanoflow dual-trap single-column liquid chromatography can quantify peptides, facilitating a 15-minute total run. The authors identified and quantified over 1000 proteins from cardiomyocytes and heterogenous aorta cells.

Plasma is one of the most widely used biological materials for diagnostic purposes [76–79]. However, its complexity and wide dynamic range hinder quantitative proteomics analysis. Therefore, several approaches have been developed to deplete highly abundant proteins and increase plasma protein coverage. Chemical and capture-based depletion methods are applied to deplete highly abundant proteins selectively. Chemical depletion depletes highly abundant proteins except albumin, and immune-affinity-based methods use antibodies targeting highly abundant proteins [80].

Regardless of the depletion strategy, the depleted proteins are highly diluted. They must be further processed before the LC-MS proteomics analysis. Although depletion helps significantly boost the number of identified proteins, the absence of depletion simplifies sample preparation, increases sample throughput, and decreases the introduction of variance and error, limiting the dynamic range. Using the DIA approach, the authors developed and optimized a dual workflow for comprehensive and fast sample preparation and analysis.

Although the dual-trap column approach and the short and steep LC gradient approach have significantly improved the analysis speed, the number of samples that can be analyzed is still low.

Using monolithic columns is a possible game changer for developing fast proteomics analysis and its application in clinics. Unlike conventionally packed separation columns, monolithic columns enable fast flow with only low back pressure, fast mass transfer of analytes, and excellent peak shape. Tomioka et al. [81] have developed and optimized a capillary flow LC-MS method for analyzing 1000 samples daily. The major challenge was optimizing the electrospray conditions to maintain high sensitivity even when using high flow rates. The authors describe the development of a high-throughput analysis, which they termed "machine-gun proteomics,". The method can be applied to analyze up to 1000 samples daily while maintaining sensitivity and the depth of proteome analysis. More specifically, the flow rate was modified, and electrospray conditions were optimized to maintain efficient ionization and spray stability at high flow rates. A sharp and short separation gradient (sub-minute) was created for the high sample throughput, and the overhead times for sample injection and the column wash and conditioning were optimized. Upon optimization, the MS sensitivity was comparable to that achieved with the nanoflow LC-MS/MS.

Changes and optimization steps for the LC combined with the separation based on ion mobility of ionized species and the liquid chromatography, minimizing LC void volume significantly expanded the separation efficiency. It helped, at least partially, overcome the trade-off between the separation speed, sensitivity of the MS, and the analysis depth.

Tomioka et al. further describe the comparison of analyzing mouse brain tissue using the newly optimized "machine gun proteomics" and the original analysis by Piehowski et al. [82]. For the initial study, three hours were needed to analyze one 100 μm tissue section compared to 1.4 minutes using the newly developed methods. The analysis speed did not negatively impact the number of identified proteins, and the analysis of 96 sections was completed in a shorter time than needed for one section using the original method.

When addressing clinical proteomics, one must remember that the number of identified proteins or peptides does not matter alone. For clinical purposes, the information extracted from samples must be quantitative, and the quantitation must be reproducible and have a low CV (Coefficient of Variation). The machine-gun approach is currently the fastest, with reportedly low CV values between 11 and 18% for 720 samples per day and 12–24% for 1000 samples per day.

Viode et al. [36] published another approach for high-throughput analysis of clinical samples, in this case plasma. The authors described a high-throughput method using preloaded tips and fast separation gradients. The method is presented as cost-effective, robust, and high-throughput compatible, including plasma depletion. The method enables in-depth profiling of plasma samples, detecting>1300 proteins per run and 60 samples per day. Depleted and tryptically digested plasma proteins were loaded onto C18 tips and directly eluted from the tip onto the separation column. The separation column used was an 8-cm long and 150 μm inner-diameter column operated with a 21-minute gradient, translating into a cycle time of 24 minutes. The authors used DDA and DIA mass spectrometric methods.

The authors used perchloric acid to deplete abundant proteins and enable the detection of viral proteins. By applying the perchloric acid depletion, it was possible to perform an untargeted detection of the SARS-CoV-2 virus in patients' plasma samples. The SARS-CoV-2 nuclear protein (NP) was identified in ~10% of analyzed samples, and the relation between the expression level of the NP and the severity of the disease. An increase in NPs abundance did not correlate with the disease's severity, but the observation frequency did.

Another application of high-throughput proteomics analysis of plasma is mapping the sepsis plasma proteome and identifying differences in patient response. Mi et al. [83] applied fast separation methods and analyzed digested plasma samples using an Evosep One LC system. Unlike in the previous study, the authors used the faster separation method, the pre-built 100 samples per day. The 100 samples per day method is executed with the 11.5 min gradient and a total cycle time of 14.4 min. Furthermore, the flow rate is higher with a 1.2 μl/min. The optimized method and instrument design maximized data acquisition time and reduced overheads, such as column equilibration, sample injection, trapping, etc. The authors analyzed plasma proteome from patients with sepsis and control groups. An impressive number of 2612 samples from 1611 were acquired from a single batch without interrupting measurements.

Sepsis is a life-threatening state where organs are dysfunctional because of a dysregulated host response to infection [84]. Currently, sepsis causes high mortality rates, and there is a lack of effective immunomodulatory therapies. Briefly, patients were divided into the six sepsis-comparator groups, and 11 proteins were found to be differentially abundant in all contrasts and all with the highest abundance in sepsis samples. Proteins found to be more abundant in sepsis were implicated in the acute-phase response (CRP, SAA1, and SAA2), the coagulation process (VWF (Von Willebrand factor), FGB (fibrinogen beta chain), and FGA), and immune or immune-regulatory functions (LBP, S100A9, FGL1 (fibrinogen-like protein 1), ORM1 (alpha-1-acid glycoprotein 1), and CD14 (cluster of differentiation 14, monocyte differentiation antigen). Furthermore, authors identified apolipoproteins, α-2-HS-glycoprotein, hepatocyte growth factor activator, plasma serine protease inhibitor (SERPINA5), TTR (transthyretin inhibited by inflammation), and transcription regulator protein BACH2 (regulates apoptosis and adaptive immunity) to have reduced abundancy in sepsis samples.

The study was extensive and combined with transcriptomics, delivering crucial knowledge for identifying putative biomarkers for sepsis. Bearing that sepsis is a life-threatening state of the organism, rapid action is needed to define and distinguish between it and other inflammation processes. The described method could be used for fast and targeted analysis of plasma sepsis and identification of the severity of the patient's state in a very short time.

3. Conclusion

In conclusion, high-throughput proteomics, in general and in clinics, is gaining momentum and is a significant step forward in technological development and improvement for fast and sensitive LC-MS/MS for comprehensive proteome analysis. The development of new stationary phases, for example, the monolithic ones and the innovative MS screening and detection methods, such as TIMS, PASEF, and DIA, enables fast sample throughput and reliable and stable qualitative and quantitative analysis of complex biological samples.

Further development is expected to follow rapidly, and proteomics will soon gain full access to clinical laboratories.

Acknowledgements

The authors acknowledge the support from Bruker and the Vinča Institute of Nuclear Sciences.

Conflict of interest

The authors declare no conflict of interest.

Appendices and nomenclature

MS	mass spectrometry
HPLC	high-performance liquid chromatography
nano-LC	nano Liquid chromatography
DDA	data-dependent acquisition
DIA	data-independent acquisition

Author details

Goran Mitulović[1*] and Jadranka Miletić Vukajlović[2]

1 Bruker Austria, Vienna, Austria

2 Vinča Institute of Nuclear Sciences, National Institute of the Republic of Serbia, University of Belgrade, Belgrade, Serbia

*Address all correspondence to: goran.mitulovic@bruker.com

References

[1] Wilhelm M, Schlegl J, Hahne H, Gholami AM, Lieberenz M, Savitski MM, et al. Mass-spectrometry-based draft of the human proteome. Nature. 2014;**509**(7502):582-587

[2] Ren AH, Diamandis EP, Kulasingam V. Uncovering the depths of the human proteome: Antibody-based Technologies for Ultrasensitive Multiplexed Protein Detection and Quantification. Molecular & Cellular Proteomics. 2021;**20**:100155

[3] Al-Amrani S, Al-Jabri Z, Al-Zaabi A, Alshekaili J, Al-Khabori M. Proteomics: Concepts and applications in human medicine. World Journal of Biological Chemistry. 2021;**12**(5):57-69

[4] Cui M, Cheng C, Zhang L. High-throughput proteomics: A methodological mini-review. Laboratory Investigation. 2022;**102**(11):1170-1181

[5] Messner CB, Demichev V, Wang Z, Hartl J, Kustatscher G, Mülleder M, et al. Mass spectrometry-based high-throughput proteomics and its role in biomedical studies and systems biology. Proteomics. 2023;**23**(7-8):e2200013

[6] Vukajlović JM, Panić-Janković T. Mass spectrometry in clinical laboratories. In: Mitulovic G, editor. Mass Spectrometry in Life Sciences and Clinical Laboratory. London, UK: IntechOpen; 2021. pp. 73-98

[7] Verrills NM. Clinical proteomics: Present and future prospects. Clinical Biochemist Reviews. 2006;**27**(2):99-116

[8] Zhang N, Liu X, Gao S, Wong CC. Parallel channels-multidimensional protein identification technology. Journal of the American Society for Mass Spectrometry. 2020;**31**(7):1440-1447

[9] Ong SE, Blagoev B, Kratchmarova I, Kristensen DB, Steen H, Pandey A, et al. Stable isotope labeling by amino acids in cell culture, SILAC, as a simple and accurate approach to expression proteomics. Molecular & Cellular Proteomics. 2002;**1**(5):376-386

[10] Aslam B, Basit M, Nisar MA, Khurshid M, Rasool MH. Proteomics: Technologies and their applications. Journal of Chromatographic Science. 2017;**55**(2):182-196

[11] Uzozie AC, Aebersold R. Advancing translational research and precision medicine with targeted proteomics. Journal of Proteomics. 2018;**189**:1-10

[12] Parker CE, Borchers CH. The special issue: Clinical proteomics for precision medicine. Proteomics – Clinical Applications. 2018;**12**(2):1600144

[13] Ye X, Cui X, Zhang L, Wu Q, Sui X, He A, et al. Combination of automated sample preparation and micro-flow LC-MS for high-throughput plasma proteomics. Clinical Proteomics. 2023;**20**(1):3

[14] Geyer PE, Holdt LM, Teupser D, Mann M. Revisiting biomarker discovery by plasma proteomics. Molecular Systems Biology. 2017;**13**(9):942

[15] Mortezai N, Harder S, Schnabel C, Moors E, Gauly M, Schlüter H, et al. Tandem affinity depletion: A combination of affinity fractionation and immunoaffinity depletion allows the detection of low-abundance components in the complex proteomes of body fluids. Journal of Proteome Research. 2010;**9**(12):6126-6134

[16] Bellei E, Bergamini S, Monari E, Fantoni LI, Cuoghi A, Ozben T, et al.

High-abundance proteins depletion for serum proteomic analysis: Concomitant removal of non-targeted proteins. Amino Acids. 2011;**40**(1):145-156

[17] Larance M, Lamond AI. Multidimensional proteomics for cell biology. Nature Reviews. Molecular Cell Biology. 2015;**16**(5):269-280

[18] Barabási AL, Oltvai ZN. Network biology: Understanding the cell's functional organization. Nature Reviews. Genetics. 2004;**5**(2):101-113

[19] Kustatscher G, Grabowski P, Schrader TA, Passmore JB, Schrader M, Rappsilber J. Co-regulation map of the human proteome enables identification of protein functions. Nature Biotechnology. 2019;**37**(11):1361-1371

[20] Lapek JD Jr, Greninger P, Morris R, Amzallag A, Pruteanu-Malinici I, Benes CH, et al. Detection of dysregulated protein-association networks by high-throughput proteomics predicts cancer vulnerabilities. Nature Biotechnology. 2017;**35**(10):983-989

[21] Wu L, Candille SI, Choi Y, Xie D, Jiang L, Li-Pook-Than J, et al. Variation and genetic control of protein abundance in humans. Nature. 2013;**499**(7456):79-82

[22] Bensimon A, Heck AJ, Aebersold R. Mass spectrometry-based proteomics and network biology. Annual Review of Biochemistry. 2012;**81**:379-405

[23] Sabidó E, Selevsek N, Aebersold R. Mass spectrometry-based proteomics for systems biology. Current Opinion in Biotechnology. 2012;**23**(4):591-597

[24] Rozanova S, Barkovits K, Nikolov M, Schmidt C, Urlaub H, Marcus K. Quantitative mass spectrometry-based proteomics: An overview. Methods in Molecular Biology. 2021;**2228**:85-116

[25] Čuklina J, Pedrioli PGA, Aebersold R. Review of batch effects prevention, diagnostics, and correction approaches. Methods in Molecular Biology. 2020;**2051**:373-387

[26] Oberg AL, Vitek O. Statistical design of quantitative mass spectrometry-based proteomic experiments. Journal of Proteome Research. 2009;**8**(5):2144-2156

[27] Krzywinski M, Altman N. Points of significance: Analysis of variance and blocking. Nature Methods. 2014;**11**(7):699-700

[28] Ransohoff DF. Bias as a threat to the validity of cancer molecular-marker research. Nature Reviews. Cancer. 2005;**5**(2):142-149

[29] Ponzini E, Santambrogio C, De Palma A, Mauri P, Tavazzi S, Grandori R. Mass spectrometry-based tear proteomics for noninvasive biomarker discovery. Mass Spectrometry Reviews. 2022;**41**(5):842-860

[30] Bruderer R, Muntel J, Müller S, Bernhardt OM, Gandhi T, Cominetti O, et al. Analysis of 1508 plasma samples by capillary-flow data-independent acquisition profiles proteomics of weight loss and maintenance. Molecular & Cellular Proteomics. 2019;**18**(6):1242-1254

[31] Sun R, Hunter C, Chen C, Ge W, Morrice N, Liang S, et al. Accelerated protein biomarker discovery from FFPE tissue samples using single-shot, short gradient microflow SWATH MS. Journal of Proteome Research. 2020;**19**(7):2732-2741

[32] Vowinckel J, Zelezniak A, Bruderer R, Mülleder M, Reiter L, Ralser M.

Cost-effective generation of precise label-free quantitative proteomes in high-throughput by microLC and data-independent acquisition. Scientific Reports. 2018;**8**(1):4346

[33] Bian Y, Bayer FP, Chang YC, Meng C, Hoefer S, Deng N, et al. Robust microflow LC-MS/MS for proteome analysis: 38 000 runs and counting. Analytical Chemistry. 2021;**93**(8):3686-3690

[34] Messner CB, Demichev V, Wendisch D, Michalick L, White M, Freiwald A, et al. Ultra-high-throughput clinical proteomics reveals classifiers of COVID-19 infection. Cell Systems. 2020;**11**(1):11-24.e4

[35] Gaun A, Lewis Hardell KN, Olsson N, O'Brien JJ, Gollapudi S, Smith M, et al. Automated 16-Plex plasma proteomics with real-time search and ion mobility mass spectrometry enables large-scale profiling in naked mole-rats and mice. Journal of Proteome Research. 2021;**20**(2):1280-1295

[36] Mc Ardle A, Binek A, Moradian A, Chazarin Orgel B, Rivas A, Washington KE, et al. Standardized workflow for precise mid- and high-throughput proteomics of blood biofluids. Clinical Chemistry. 2022;**68**(3):450-460

[37] Zelezniak A, Vowinckel J, Capuano F, Messner CB, Demichev V, Polowsky N, et al. Machine learning predicts the yeast metabolome from the quantitative proteome of kinase knockouts. Cell Systems. 2018;**7**(3):269-83.e6

[38] Desmet G, de Beeck JO, Van Raemdonck G, Van Mol K, Claerebout B, Van Landuyt N, et al. Separation efficiency kinetics of capillary flow micro-pillar array columns for liquid chromatography. Journal of Chromatography. A. 2020;**1626**:461279

[39] Hayes R, Ahmed A, Edge T, Zhang H. Core-shell particles: Preparation, fundamentals and applications in high performance liquid chromatography. Journal of Chromatography. A. 2014;**1357**:36-52

[40] Corbett JR, Robinson DE, Patrie SM. Robustness and ruggedness of isoelectric focusing and superficially porous liquid chromatography with Fourier transform mass spectrometry. Journal of the American Society for Mass Spectrometry. 2021;**32**(1):346-354

[41] Cupp-Sutton KA, Wu S. High-throughput quantitative top-down proteomics. Molecular Omics. 2020;**16**(2):91-99

[42] Suhre K, McCarthy MI, Schwenk JM. Genetics meets proteomics: Perspectives for large population-based studies. Nature Reviews. Genetics. 2021;**22**(1):19-37

[43] CAN NÖ. Development of validated and stability-indicating LC-DAD and LC-MS/MS methods for determination of avanafil in pharmaceutical preparations and identification of a novel degradation product by LCMS-IT-TOF. Molecules. 2018;**23**(7):1771

[44] Ctortecka C et al. Automated single-cell proteomics providing sufficient proteome depth to study complex biology beyond cell type classifications. Nature Communications. 2024;**15**(1):5707

[45] Moore JL et al. Prospective on imaging mass spectrometry in clinical Diagnostics. Molecular & Cellular Proteomics. 2023;**22**(9):100576

[46] Demichev V, Szyrwiel L, Yu F, Teo GC, Rosenberger G, Niewienda A, et al. Dia-PASEF data analysis using FragPipe and DIA-NN for deep proteomics of low sample amounts. Nature Communications. 2022;**13**(1):3944

[47] Lv J, Shan X, Yang H, Wen Y, Zhang X, Chen H, et al. Single cell proteomics profiling reveals that embryo-secreted TNF-α plays a critical role during embryo implantation to the endometrium. Reproductive Sciences. 2022;**29**(5):1608-1617

[48] Meier F, Brunner AD, Frank M, Ha A, Bludau I, Voytik E, et al. diaPASEF: Parallel accumulation-serial fragmentation combined with data-independent acquisition. Nature Methods. 2020;**17**(12):1229-1236. DOI: 10.1038/s41592-020-00998-0

[49] Demichev V, Messner CB, Vernardis SI, Lilley KS, Ralser M. DIA-NN: Neural networks and interference correction enable deep proteome coverage in high throughput. Nature Methods. 2020;**17**(1):41-44

[50] Wen C, Wu X, Lin G, Yan W, Gan G, Xu X, et al. Evaluation of DDA library-free strategies for Phosphoproteomics and Ubiquitinomics data-independent acquisition data. Journal of Proteome Research. 2023;**22**(7):2232-2245

[51] Krasny L, Huang PH. Data-independent acquisition mass spectrometry (DIA-MS) for proteomic applications in oncology. Molecular Omics. 2021;**17**(1):29-42

[52] Isaksson M, Karlsson C, Laurell T, Kirkeby A, Heusel M. MSLibrarian: Optimized predicted spectral libraries for data-independent acquisition proteomics. Journal of Proteome Research. 2022;**21**(2):535-546

[53] Kelly V et al. Low cell number proteomic analysis using in-cell protease digests reveals a robust signature for cell cycle state classification. Molecular & Cellular Proteomics. 2022;**21**(1):1-18

[54] Derks J et al. Increasing the throughput of sensitive proteomics by plexDIA. Nature Biotechnology. 2023;**41**(1):50-59

[55] Liu Y, Sun K, Shao C, Shi X, Zeng J, Guo R, et al. Performance of nanoflow liquid chromatography using core-shell particles: A comparison study. Journal of Chromatography A. 2021;**1648**:462218

[56] Meier F, Park MA, Mann M. Trapped ion mobility spectrometry and parallel accumulation-serial fragmentation in proteomics. Molecular & Cellular Proteomics: MCP. 2021;**20**:100138

[57] Perkel JM. Single-cell proteomics takes Centre stage. Nature. 2021;**597**(7877):580-582

[58] Stejskal K, Op de Beeck J, Dürnberger G, Jacobs P, Mechtler K. Ultrasensitive NanoLC-MS of Subnanogram protein samples using second generation micropillar Array LC technology with Orbitrap Exploris 480 and FAIMS PRO. Analytical Chemistry. 2021;**93**(25):8704-8710

[59] Bader JM, Albrecht V, Mann M. MS-based proteomics of body fluids: The end of the beginning. Molecular & Cellular Proteomics. 2023;**22**(7)

[60] Bian Y, Zheng R, Bayer F, Wong C, Chang YC, Meng C, et al. Robust, reproducible and quantitative analysis of thousands of proteomes by micro-flow LC-MS/MS. Nature Communications. 2020;**11**:157

[61] Messner C, Demichev V, Bloomfield N, Yu J, White M, Kreidl M, et al. Ultra-fast proteomics with scanning SWATH. Nature Biotechnology. 2021;**39**:846

[62] Szyrwiel L, Gille C, Mülleder M, Demichev V, Ralser M. Speedy-PASEF: Analytical flow rate chromatography and trapped ion mobility for deep

high-throughput proteomics. bioRxiv. 2023:2023.02.17.528968.

[63] Percy AJ et al. Multiplexed MRM-based assays for the quantitation of proteins in mouse plasma and heart tissue. PROTEOMICS. 2017;**17**(7):1600097

[64] Percy AJ et al. Multiplexed panel of precisely quantified salivary proteins for biomarker assessment. PROTEOMICS. 2017;**17**(6):1600230

[65] Muntel J, Gandhi T, Verbeke L, Bernhardt OM, Treiber T, Bruderer R, et al. Surpassing 10 000 identified and quantified proteins in a single run by optimizing current LC-MS instrumentation and data analysis strategy. Molecular Omics. 2019;**15**(5):348-360

[66] Suckau D, Evers W, Belau E, Pengelley S, Resemann A, Tang W, et al. Use of PASEF for accelerated protein sequence confirmation and De novo sequencing with high data quality. Methods in Molecular Biology (Clifton, NJ). 2022;**2313**:207-217

[67] Aballo TJ, Roberts DS, Melby JA, Buck KM, Brown KA, Ge Y. Ultrafast and reproducible proteomics from small amounts of heart tissue enabled by azo and timsTOF pro. Journal of Proteome Research. 2021;**20**(8):4203-4211

[68] Vasilopoulou CG, Sulek K, Brunner AD, Meitei NS, Schweiger-Hufnagel U, Meyer SW, et al. Trapped ion mobility spectrometry and PASEF enable in-depth lipidomics from minimal sample amounts. Nature Communications. 2020;**11**(1):331

[69] Guergues J, Wohlfahrt J, Stevens SM Jr. Enhancement of proteome coverage by ion mobility fractionation coupled to PASEF on a TIMS-QTOF instrument. Journal of Proteome Research. 2022;**21**(8):2036-2044

[70] Sharaf BM, Giddey AD, Al-Hroub HM, Menon V, Okendo J, El-Awady R, et al. Mass spectroscopy-based proteomics and metabolomics analysis of triple-positive breast cancer cells treated with tamoxifen and/or trastuzumab. Cancer Chemotherapy and Pharmacology. 2022;**90**(6):467-488

[71] Spanier B, Laurençon A, Weiser A, Pujol N, Omi S, Barsch A, et al. Comparison of lipidome profiles of Caenorhabditis elegans-results from an inter-laboratory ring trial. Metabolomics. 2021;**17**(3):25

[72] Lenčo J, Vajrychová M, Pimková K, Prokšová M, Benková M, Klimentová J, et al. Conventional-flow liquid chromatography–mass spectrometry for exploratory bottom-up proteomic analyses. Analytical Chemistry. 2018;**90**(8):5381-5389

[73] Kaspar-Schoenefeld S, Krieger JR, Martelli C, König A-C, Hauck S, Johansson S, et al. High-throughput proteome profiling with low variation in a multi-center study using dia-PASEF. bioRxiv. 2024:2024.05.29.596405

[74] Kreimer S, Haghani A, Binek A, Hauspurg A, Seyedmohammad S, Rivas A, et al. Parallelization with dual-trap single-column configuration maximizes throughput of proteomic analysis. Analytical Chemistry. 2022;**94**(36):12452-12460

[75] Kreimer S, Binek A, Chazarin B, Cho JH, Haghani A, Hutton A, et al. High throughput proteomic analysis of heterogeneous cell populations with NanoDTSC. bioRxiv. 2023:2023.01.06.522908

[76] Tuck MK, Chan DW, Chia D, Godwin AK, Grizzle WE, Krueger KE, et al. Standard operating procedures for serum and plasma collection: Early detection research network consensus statement standard operating procedure integration working group. Journal of Proteome Research. 2009;**8**(1):113

[77] Miletić J, Drakulić D, Pejić S, Petković M, Ilić TV, Miljković M, et al. Prooxidant-antioxidant balance, advanced oxidation protein products and lipid peroxidation in Serbian patients with Parkinson's disease. The International Journal of Neuroscience. 2018;**128**(7):600-607

[78] Grigorov I, Pejić S, Todorović A, Drakulić D, Veljković F, Vukajlović JM, et al. Serum high-mobility group box 1 and Heme Oxygenase-1 as biomarkers in COVID-19 patients at hospital admission. International Journal of Molecular Sciences. 2023;**24**(17)

[79] Miletić Vukajlović J, Drakulić D, Pejić S, Ilić TV, Stefanović A, Petković M, et al. Increased plasma phosphatidylcholine/lysophosphatidylcholine ratios in patients with Parkinson's disease. Rapid Communications in Mass Spectrometry. 2020;**34**(4):e8595

[80] Millioni R, Tolin S, Puricelli L, Sbrignadello S, Fadini GP, Tessari P, et al. High abundance proteins depletion vs low abundance proteins enrichment: Comparison of methods to reduce the plasma proteome complexity. PLoS One. 2011;**6**(5):e19603

[81] Tomioka A, Tomioka R, Mori I, Isobe Y, Arita M, Imami K, et al. One thousand samples per day capillary-flow LC/MS/MS for high-speed, high-sensitivity and in-depth proteomics. bioRxiv. 2023:2023.06.05.543682

[82] Piehowski PD, Zhu Y, Bramer LM, Stratton KG, Zhao R, Orton DJ, et al. Automated mass spectrometry imaging of over 2000 proteins from tissue sections at 100-μm spatial resolution. Nature Communications. 2020;**11**(1):8

[83] Mi Y, Burnham KL, Charles PD, Heilig R, Vendrell I, Whalley J, et al. High-throughput mass spectrometry maps the sepsis plasma proteome and differences in patient response. Science Translational Medicine. 2024;**16**(750):eadh0185

[84] Singer M, Deutschman CS, Seymour CW, Shankar-Hari M, Annane D, Bauer M, et al. The third international consensus definitions for sepsis and septic shock (Sepsis-3). Journal of the American Medical Association. 2016;**315**(8):801-810

Chapter 7

Metabolomic Diagnostic in Inherited Metabolic Disorders: Historical Progress and Future Perspectives

Clara Oliva Mussarra and Judit García-Villoria

Abstract

Inherited metabolic disorders (IMD) encompass a diverse range of monogenic disorders disrupting metabolic pathways, leading to significant morbidity and mortality. While some treatable IMD are part of newborn screening programs, the majority remain challenging to diagnose early. Targeted metabolomics, analyzing specific metabolites in biological fluids, has evolved from disease-specific tests to advanced chromatographic techniques such as gas chromatography, high performance liquid chromatography and GC-mass spectrometry (MS). The development of tandem MS in the 1990s marked a pivotal advancement, allowing the detection of various metabolites in a single analysis. However, current metabolic diagnostic methods still involve manual, time-intensive processes using multiple analyzers, limiting their widespread application. Diagnosis often hinges on recognizing nonspecific clinical symptoms, risking missed cases without appropriate metabolic testing needed to achieve the diagnosis. To address these challenges, there is a pressing need for untargeted metabolomics—a holistic approach using high-resolution MS to measure thousands of metabolites in a single analysis. This approach promises to revolutionize diagnostics by enabling comprehensive metabolite detection, optimizing resources, and streamlining diagnostic workflows. Integrating metabolomics with other omics technologies, especially genomics, is essential to achieving definitive IMD diagnoses. This multidisciplinary approach not only enhances IMD diagnosis but also supports personalized medicine.

Keywords: inherited metabolic disorders, tandem mass spectrometry, high resolution mass spectrometry, metabolomics, high performance chromatography

1. Introduction

1.1 Inherited metabolic disorders (IMD)

Inherited metabolic disorders (IMD), traditionally known as inborn errors of metabolism (IEM), are a group of rare disorders caused by mutations in genes

IntechOpen

encoding enzymes, membrane transporters, or other functional protein that affect metabolic pathways. As a result, dysfunctions occur in the biosynthesis, utilization, degradation, or storage of biochemical substrates such as amino acids, carbohydrates, lipids, cofactors, or complex macromolecules. These alterations can lead to a deficit in energy production or a pathological accumulation of toxic metabolites or macromolecules, often accompanied by a deficit of essential products [1].

The first IMD was described in 1859 when Dr. Boedker identified the chemical responsible for darkening urine in a group of patients and named it "alkapton" [2]. The unifying term of IEM was described in 1908 by Sir Archibald Garrod in reference to alkaptonuria, pentosuria, cystinuria and albinism. Since then, many laboratory tests have been developed over the years for the diagnosis of this group of diseases, and the clinical laboratory has played a key role both in the description of various IEM and in their diagnosis.

The latest international classification of IMD includes 1450 different disorders [3]. The individual prevalence of each IMD is between 1 and 10 per 100,000 live births, which is why they are considered rare diseases. However, collectively, they have an overall prevalence of approximately 1 in 2000 live births, representing a significant public health issue [4], being a diverse array of disorders that encompass a significant amount of morbidity and mortality worldwide [1]. Additionally, given the large number of entities described in recent years, it is expected that this prevalence is even higher, especially considering the underdiagnosis associated with many of these diseases [5, 6].

The age of onset and clinical severity of IMD vary significantly from one condition to another, ranging from extremely severe forms, including prenatal presentation, to mild adult-onset forms, and even some asymptomatic deficiencies. Additionally, these diseases exhibit a wide range of clinical manifestations that can affect any organ or system, varying according to the type of disorder and the severity of the mutation. Although most IMD have an autosomal recessive inheritance pattern, there are also deficiencies with autosomal dominant or X-linked inheritance. Additionally, there are some diseases caused by alterations in mitochondrial DNA that are inherited maternally [6].

While some IMD present characteristic symptoms that allow for a high index of suspicion, many others present common and nonspecific symptoms that complicate their identification and can lead to significant diagnostic delays [5].

The treatment of IMD depends on their specific type. Although there is no cure for many of these conditions, some have available therapies that help improve symptoms and the quality of life for patients. In some cases, the available treatments can drastically improve the natural course of the disease and the clinical outcomes for patients. Treatment options include special diets with substrate restriction or supplementation with products and/or enzymatic cofactors, drugs, enzyme replacement therapy, gene therapy, or organ transplantation. Obtaining a rapid and accurate diagnosis for these treatable disorders is essential to avoid irreversible sequelae [7, 8].

One of the most distinctive aspects of IMD is the existence of biochemical phenotypes for many of them because of the underlying metabolic alterations. These can be identified through the analysis of specific biomarkers that result from the accumulation or depletion of metabolic intermediates and products in biological fluids. Additionally, for many diseases caused by enzyme deficiencies, it is also possible to measure enzymatic activity to confirm their diagnosis. Various functional studies can also be performed for other types of proteins, such as those involved in cellular transport or uptake [4, 7].

Although in some cases the biochemical phenotype may be sufficient to confirm the diagnosis, genetic studies are often necessary to provide genetic counseling to the family. In other cases where specific biomarkers do not exist, molecular studies are also necessary to reach a diagnosis. The existence of a biochemical phenotype originally defined this group of diseases. However, the implementation of next generation sequencing (NGS) has led to the description of many entities that, despite affecting different metabolic pathways, do not present identifiable biochemical alterations through conventional diagnostic tests [6, 9].

1.2 Diagnostic techniques for IMD

The classical diagnostic sequence for clinical suspicion of IMD begins with the analysis of a limited number of metabolites in biological fluids. This is followed by the study of potentially altered proteins and is then confirmed by genetic studies. However, this sequence is not always valid, either because a particular IMD does not have specific metabolites or biomarkers, or because the number of tests to be performed is very high. The selection of diagnostic tests is based on clinical phenotype; however, the clinical symptoms of IMD are often nonspecific, making diagnosis difficult.

Over the years, there has been an evolution in how these diseases are diagnosed, leading to continuous change and evolution in clinical laboratories for IMD diagnostics, see **Figure 1**. This evolution encompasses advancements not only in biochemical tests but also in genomic studies. Consequently, a reference laboratory for the diagnosis of IMD must have access to various instrumental platforms to achieve accurate diagnoses of these conditions [10].

Focusing on the analysis of metabolites, initially, diagnostic laboratories for IMD conducted specific tests manually for the diagnosis of known diseases. With advancements in chromatographic technology, laboratories have developed tests capable of detecting different metabolites in a single test to diagnose various conditions. The

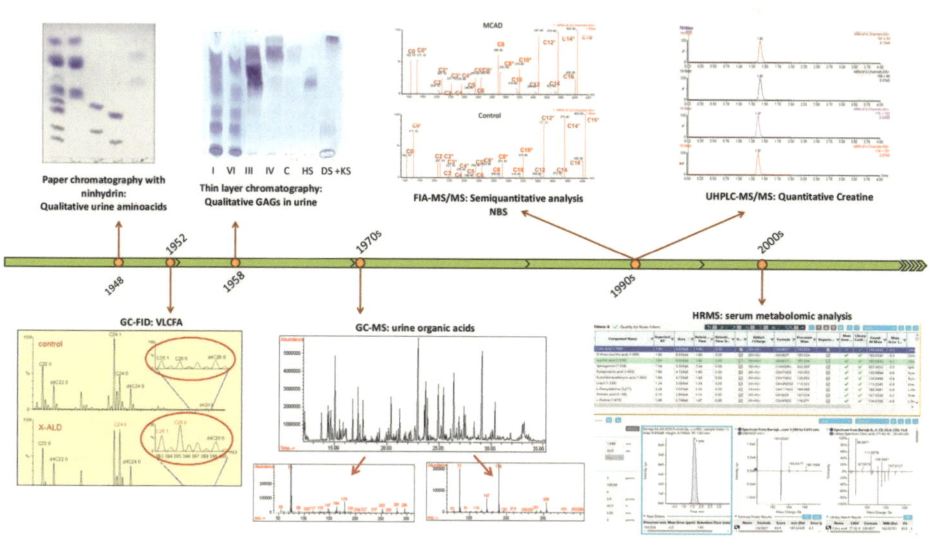

Figure 1.
Evolution of different chromatographic and mass spectrometry techniques throughout the history of inherited metabolic diseases diagnosis. FIA-MS/MS: flow injection analysis in tandem mass spectrometry. GAGs: glycosaminoglycans. GC-FID: gas chromatography with flame ionization detector. GC-MS: gas chromatography with mass spectrometry detector. MS/MS: ultra-high performance liquid chromatography with tandem mass spectrometry. NBS: newborn screening. UHPLC-HRMS: high-resolution mass spectrometry. VLCFA: very long chain fatty acids.

implementation of chromatography with tandem mass spectrometry (MS/MS) was a turning point in the detection of IMD, as it allowed for the detection of many metabolites from different metabolic pathways [10].

Currently, reference laboratories for IMD diagnostics can analyze different types of samples such as urine, plasma, dried blood spots (DBS), lymphocytes, granulocytes, and others using various chromatographic systems—gas chromatography (GC) coupled or not with mass spectrometry (MS), and liquid chromatography (LC) coupled or not with tandem mass spectrometry (MS/MS).

We are now reaching maximum optimization with the goal of achieving the minimum number of tests for diagnosing the maximum number of diseases. In this line, the application of high-resolution mass spectrometry (HRMS) now allows the diagnosis of around 500 IMDs in a single test.

On the other hand, enzymatic analyses are conducted using specific tests employing various spectrophotometric and MS/MS techniques in serum, leukocytes, lymphocytes, dried blood spots, skin fibroblasts, or muscle biopsies, depending on the disorder. These methods are predominantly used for diagnosing lysosomal and mitochondrial diseases [11].

Finally, confirmation of the diagnosis is achieved through genetic studies. If biochemical studies suggest a disease caused by a specific gene, Sanger sequencing of the corresponding gene is performed. However, when the biochemical phenotype suggests conditions caused by mutations in different genes, it is more cost-effective to analyze them using next-generation sequencing (NGS) panels or whole exome sequencing (WES), which are now included in diagnostic routines [7, 9].

Advancements in sequencing technology and cost reductions are progressively placing genetic tests at the forefront of diagnostics. Nonetheless, biochemical analyses remain essential as they provide functional information crucial for interpreting genetic variants of uncertain clinical significance (VUS). The complementarity of biochemical and molecular studies is pivotal in maximizing diagnostic accuracy. Additionally, biochemical markers remain particularly important for patient monitoring and evaluating responses to therapies [9, 12].

It is important to emphasize the significance of introducing quality management systems in diagnostic laboratories through the incorporation of ISO standards. ISO standards were first described in 1980, and the establishment of quality standards for clinical laboratories occurred with the publication of the Clinical Laboratory Improvement Amendments (CLIA) in 1988 [13]. In pursuit of this goal, the European Research Network for evaluation and improvement of screening, Diagnosis, and treatment of Inherited disorders of Metabolism (ERNDIM) was founded in 1994 to provide an external quality control program for laboratory techniques used in IMD diagnostic [14]. These advancements were crucial for gradually enhancing result quality and fostering harmonization among laboratories [15].

This chapter will describe the role that chromatography has played over the years in the diagnosis and study of IMD, including their evolution and emerging pathways.

2. Progress in metabolomic studies for IMD diagnosis

2.1 Before chromatography

The first biochemical analysis for the detection of an IMD was described in the 1930s, when Fölling used a simple chemical test to screen the urine of two mentally retarded siblings who had a peculiar odor in their urine. The urine reacted with a solution of ferric chloride, producing a green color. Later, they identified the substance as phenylpyruvic acid. This discovery allowed him to diagnose more cases of phenylketonuria (PKU) by screening mentally retarded patients in institutions. The application of other simple tests such as Benedict's for reducing sugars, cyanide-nitroprusside test for disulfides (used for cystinuria) and Berry's spot test for mucopolysaccharides, along with the invention of spectrophotometry by Arnold J. Beckman in 1940, preceded the development and popularization of chromatographic techniques [16, 17].

In 1960, Robert Guthrie's pioneering work in developing a bacterial inhibition assay to measure phenylalanine in DBS provided the means for whole-population screening to detect PKU. Its implementation started in Massachusetts (USA) in 1963, and in the following decades, it became established in developed countries [18]. This test contributed to popularizing the use of a filter paper card to collect a small drop of blood from a baby's heel, allowing for easy transport to a testing laboratory. However, over the years, this method was gradually replaced by chromatographic techniques (such as thin layer chromatography), and later on, MS/MS methods [19, 20].

2.2 First chromatographic tests

Over the 20th century, there was a significant evolution in the diagnosis of IMD thanks to the advances in chromatographic techniques. The invention of new technologies for detecting different metabolites led to the publication of various methods throughout the century. The first procedures described were using paper chromatography (1944), followed by GC (1952), thin layer chromatography and high-performance liquid chromatography (HPLC) in 1962. The detectors used in these chromatographic techniques were relatively nonspecific, and the identification of metabolites was based on their retention time [10].

For example, in the diagnosis of defects in amino acids metabolism, different chromatographic techniques have been described. In 1948, Stein and Moore described the photometric ninhydrin method, a quantitative method that took 2 weeks. This method was then improved in 1958 by Spackman et al., who reduced the time to 24 hours by adding a hydrolase. In the same year, the first commercial quantitative amino acid analyzer was developed using ion exchange resin chromatography and a visible wavelength detector allowing results to be obtained in less than 12 hours [21].

Other technologies apart from chromatography were developed, such as radioimmunoassay and enzyme-immunoassays for measuring urine and plasma carnitine and acylcarnitine concentrations. In 1986, a method was developed to differentiate between individual acylcarnitines by exchanging radioactive carnitine to isotopic equilibrium and separating acylcarnitines using HPLC. This method has been used to detect short-chain acylcarnitines on thin-layer chromatograms, paper chromatograms, and column effluents [22].

2.3 Gas chromatography-mass spectrometry (GC-MS)

A first breakthrough was the outcome of determination of organic acids with GC-MS [23]. The addition of mass spectrometry to GC applications dramatically improved the analysis of organic acids by providing mass spectral identification of each compound at a particular retention time. This combined technique, first published by Thompson et al. [24], became a gold standard for identification of metabolic disorders from urine specimens. The detection of multiple metabolites in a one-hour run not only expanded the number of the patients analyzed but also allowed the identification of new diseases. For example, it allowed to differentiate between propionic aciduria and methylmalonic aciduria, since then classified both as ketosis hyperglycinaemia [23].

Prior to GC-MS, there were already different organic acidurias described, such as propionic aciduria [25], methylmalonic aciduria [26, 27], isovaleric aciduria [28], marple syrupe urine disease [29] and Glutaric aciduria type II [30], but the procedures for the identification of these diseases required isolation of the characteristic volatile acids from blood or urine by vacuum distillation, extraction or silicic acid absorption.

2.4 Tandem mass spectrometry (MS/MS)

The next breakthrough in IMD laboratory diagnostic was in the late 20th century with the advent of MS/MS [31]. This groundbreaking technology revolutionized the field of clinical diagnosis. Unlike the previous methods that were time-consuming, lacked specificity and only allowed the determination of few metabolites in one test, MS/MS would allow the development of very specific, sensitive and robust methods that require few samples volume and enables the identification of more than 40 IMD in a single sample, which was key for its application in dried blood spots paved the way toward expanded newborn screening using biochemical markers [19, 20].

This approach led to the definition of the term metabolomics in the 90s to describe techniques aimed at measuring the metabolites present within a cell, tissue or organism during a genetic alteration or physiological stimulus [32].

The flow injection analysis (FIA), that allows injecting samples directly into the mass spectrometer, has been widely used in newborn screening programs as these programs need to analyze a large number of samples in a very short period of time. However, this technique has certain limitations, such as difficulty in differentiating isomers and the use of a single calibration point with an internal standard, which can introduce certain constraints [31, 33].

Later on, HLPC was coupled to MS/MS improving the specific detection of metabolites and allowing chromatographic separation of the compounds. Technological advances in column development for LC, such as the reduction in column particle size and the use of superficially porous particles, allowed for increased back pressure and more efficient separation with shorter analysis times. This new column technology for LC-MS/MS, known as ultra-high-performance liquid chromatography (UHPLC), emerged alongside triple quadrupole detectors, significantly reducing turnaround time and enabling the analysis of many more samples each day [34, 35].

Thus, early years of the 21st century have served to establish multiple methods using HPLC or UPLC-MS/MS for the analysis of multiple targeted metabolites for the diagnosis and treatment monitoring of various IMD [36]. At first, sample processes with derivatization were the most widely used. Later, with the emergence of new MS/MS with greater sensitivity and rapid polarity changes from negative to positive ionization, it became possible to assess compounds without the need for derivatization, simplifying the process, and allowing the analysis of compounds regardless of their ionization condition.

In recent years, with this technology, there has been a growing trend to analyze broader panels of metabolites in a single sample, but it presents limitations in terms of the number of metabolites, and they must always be previously known [37, 38].

However, with the incorporation of high-resolution mass spectrometry (HRMS) these limitations can be overcome, as it is possible to analyze thousands of metabolites in a single sample and in a single analysis that do not have to be predefined. This new approach is known as untargeted metabolomics. The analysis of data obtained by this technology can be managed in a targeted or non-targeted manner, offering seemingly infinite possibilities not only to discover new biomarkers or diseases but also to apply in diagnostics, ushering in a new era of "next-generation metabolic screening" [39, 40].

There are certain groups of IMD where HRMS is already well established in clinical diagnosis, such as in congenital disorders of glycosylation (CDG). These disorders are IMD caused by abnormalities in protein and/or lipid glycosylation, presenting with a highly variable clinical phenotype. When CDG is suspected, an initial screening is typically performed through sialotransferrin analysis in serum by HPLC o capillary electrophoresis. If the screening is positive, various HRMS techniques are then used to precisely identify the enzymatic or transporter defect by analyzing glycan structures, a field known as glycomics [41, 42].

The advancement in the development of various chromatography and mass spectrometry tests has been accompanied by progress in genetics. The culmination of all these advancements is the integration of omics into clinical laboratories, either through their separate use or the integration of some or all of them [10].

2.5 Targeted vs. untargeted metabolomics

For decades, targeted metabolomics methods have been widely and successfully used for the diagnosis and monitoring of patients with IMD [36]. However, these methods are more clinician-dependent, and clinical symptomatology typically guides the selection of specific analyses for an individual patient. Thus, this strategy heavily relies on the completeness of clinical information and therefore carries the risk of false negatives if a metabolic test has not been performed due to nonspecific clinical presentation or incomplete description of patient symptoms [39].

Targeted metabolomics requires knowledge of the metabolites to be detected and is based on metabolite-specific signal [43]. Thus, it measures a pre-defined group of biochemically characterized and interpreted metabolites, which is a subset of the metabolome. The metabolites analyzed are already known and so is their biochemical pathway, which impedes the discovery of novel metabolic perturbations. The positive side is that using isotopically labeled standards, the metabolites can be quantified with high quality standards [44–47].

On the other hand, untargeted metabolomics has significantly matured in the last decade and offers improved diagnostic performance compared to traditional approaches [39, 48–51]. It involves global profiling of the metabolome and allows for the discovery of new biomarkers. Nowadays, it is beginning to change the diagnostic approach of IMD due to its broader coverage and scope, and it is increasingly being used in the investigation and diagnosis of patients with IMD [52], often referred to as next-generation metabolic screening.

In untargeted metabolomics studies, achieving a complete and unbiased detection of all metabolites within a single sample is currently not feasible due to the heterogenous chemical properties of physiological metabolites and their wide range of concentrations, from picomolar to millimolar [53–55]. However, with appropriate validation to understand the metabolite coverage of an assay, untargeted metabolomics platforms have proven to be powerful tools for discovery and diagnostic [56, 57].

The most widely used mass spectrometers for UHPLC–HRMS are Orbitrap and quadrupole-time of flight-based systems (QTOF), both offering high mass accuracy and resolutions. Generally speaking, Orbitrap has higher resolution, while QTOF has a broader mass range and faster acquisition speed [58, 59]. Next-generation metabolic technology can be used in a completely untargeted manner to detect all the metabolites present in a sample. Bioinformatics pipelines then process all the information, which is time-consuming and requires complex computational resources. Alternatively, a targeted approach can be used for post-analytical processing, where bioinformatics pipelines analyze specific metabolites of interest within all the generated data [39].

In the last decade, this targeted approach with HRMS has been implemented in clinical laboratories. It enables the analysis of the entire metabolome related to IMD using a simple process of the sample without the need for derivatization. In a single analysis, it is possible to analyze thousands of metabolites with similar polarity, although both positive and negative ionization modes must be analyzed separately for their detection in HRMS. In any case, the use of HRMS improves the traditional approach, which required multiple analyses per sample for diagnosis. Moreover, the application of HRMS technology is not subject to the selection of tests based on clinical presentation, allowing the study of the same metabolites analyzed by conventional targeted metabolomics tests for the diagnosis of IMD. This reduces the likelihood of false negatives and shortens the diagnostic time. An example of this approach is described by Coene et al. [39], where they highlight the advantages of combining targeted and untargeted post-analytical pipeline. They propose to obtain data in an untargeted manner and then, depending on the pipeline used for data processing, to perform either targeted or untargeted analyses. This allows for the diagnosis of known IMD through targeted processing, while also identifying novel biomarkers and diseases with untargeted analysis. It is important to emphasize that all the analysis performed by Coene et al. were performed under ISO:15189 accreditation for medical laboratories [48], ensuring that this type of analysis meets adequate quality standards with reproducibility of the results and reliable data processing. These examples illustrate how clinical laboratories are increasingly transitioning to the use of HRMS for the diagnosis of IMD as it offers several advantages.

A similar approach is described by Steinbusch et al. for urine samples [60], using a single high-resolution UHPLC-QTOF platform. They tested 78 IMD and found the correct diagnosis in 68 of them. In 7 cases, diagnosis could not be determined because the patients were under treatment and had a normal metabolite profile, and the remaining 3 diseases were related to creatine metabolism (AGAT deficiency, GAMT deficiency, and creatine transporter deficiency), where technical limitations arose as creatine exhibited high within-run variation, making differentiation between patients and control groups challenging. Other studies have also demonstrated that untargeted metabolomics can improve IMD screening, especially in cases where VUS and false positive/negative results were obtained with classical biomarkers. For instance, Sebaa et al. studied patients with MCAD deficiency (MCADD) and VLCAD deficiency (VLCADD) and demonstrated that metabolomics could detect thousands of altered metabolites within a very small number of biological samples, thus reflecting the pathological status of the disease. They identified some up- and down-regulated endogenous metabolites that made it possible to differentiate patients with MCADD or VLCADD from control groups, providing a functional impact of the identified variants and allowing their pathogenicity to be demonstrated. Additionally, they discovered new biomarkers for these diseases that have yet to be validated [61, 62].

A similar study was performed by the groups of Jacob et al. [63] and Miller et al. [51] using samples of glutaric aciduria type I (GA-1) patients. They identified several significantly dysregulated metabolites in patients compared to the control group. Moreover, Sabi et al. found potential biomarkers capable of predicting newborns with transient elevation of glutaryl carnitine (C5DC) in the NBS process, which could help avoid falsely suspected cases of GA-1. However, further studies are needed to validate their findings.

In summary, untargeted metabolomics has a great potential in novel biomarkers discovery and characterizing diseases, and for many years this technology has been applied for this purpose, especially in the field of research. In the last decade, HRMS has been emerging in the diagnosis of IMDs. However, while untargeted metabolomics provides more information than targeted metabolomics, targeted metabolomics remains as the best option for quantification [40].

3. Future perspectives in biochemical diagnosis in IMD

3.1 Integration of HRMS and omics

The term "metabolome" was introduced in 1998, analogous to terms like genome, transcriptome, and proteome. It formally refers to the complete set of small molecules present in a cell or organism [64]. Exploring the metabolome of patients with IMD could expand the understanding of personalized medicine for these disorders. Additionally, in contrast to current diagnosis using targeted evaluation of metabolites, untargeted metabolomics offers the potential to consolidate multiple targeted assays into a single platform. In addition, this approach holds promise for accelerating discovery by identifying novel biomarkers and detecting previously unknown IMD [59].

Advances in omics technologies have facilitated the study of complex diseases by exploring the relationships between the genome, transcriptome, proteome, and metabolome. This integrated approach enhances our understanding of associated phenotypes and accelerates progress toward personalized medicine [12, 65].

However, while genomes, transcriptomes and proteomes can be mapped with high comprehensiveness, metabolome coverage for most organisms remains relatively limited. The structural complexity of natural metabolites, combined with their wide concentration range in living organisms (from sub-nanomolar to millimolar), presents significant challenges in developing a single analytical platform for the analysis of the whole metabolome [58]. Currently, HRMS provides the most extensive coverage of the metabolome compared to other widely used techniques and has been adopted by many laboratories for the study and diagnosis of IMD [58, 59].

Comparing results across different platforms is challenging because there is limited understanding of how data from various analytical techniques can be effectively compared, despite some efforts in this direction. Moreover, another disadvantage is that only approximately 65% of metabolites are quantifiable in all three body fluids (plasma, urine, and cerebrospinal fluid). This underscores the importance of selecting the appropriate biofluid depending on the specific focus of the study [66].

Nevertheless, there are already some examples of how integrating "-omics" data serves as a valuable tool for diagnosing and discovering IMD. This integration can aid in prioritizing variants identified through whole exome sequencing (WES) by providing a hypothesis-driven approach. But can also support the functional analysis of variants of unknown significance, thereby enhancing our understanding and interpretation of genetic findings in these disorders [67]. Thus, untargeted metabolomic analysis can contribute significantly to variant interpretation [9, 66, 68–71].

In their "cross-omics" study, Kerkhofs et al. showed that for accurate prioritization of disease-causing genes in IMD, it is essential to consider not only the primary pathway of the affected protein but also the broader network of metabolites [68]. Alaimo et al. reviewed data from 170 patients who underwent both WES and untargeted metabolomics [69]. They found that metabolomic data contributed to variant interpretation in 74 individuals (43.5%), and in 21 of these cases (12.3%), it confirmed a clinical diagnosis. Tang et al., also obtained similar results, and demonstrated that combining metabolomic screening with genetic screening improved the sensitivity of genetic diagnosis [72]. The American College of Medical Genetics and Genomics (ACMG) criteria give provided specific guidelines provision for incorporating results from high-quality functional validation studies into variant classification [73, 74]. Therefore, the integration of "-omic" technologies will provide the best approach for re-evaluating previously unsolved cases [36].

A different approach in "-omics" integration is exemplified by the discovery of NANS deficiency in patients with intellectual disability, dysmorphisms, and skeletal dysplasia. Argov et al. performed WES and metabolomic study in a patient, where untargeted metabolomics detected elevated levels of the metabolite N-acetylmannosamine, which could only be caused by two enzymatic deficiencies in the de novo sialic acid synthesis pathway. Integrating these datasets facilitated the identification of the disease gene (NANS) and validated the deleterious impact of recessive variants through the accumulation of N-acetylmannosamine, which also served as a new biomarker leading to subsequent diagnoses worldwide [75, 76]. So, in this example, "-omics" integration enabled both patient diagnosis and biomarker discovery.

There are also other examples where "-omics" has proven to be a powerful tool in discovering and diagnosing CDGs. The integration of genomics and glycomics has led to the discovery of previously unknown CDGs, such as CCDC115-CDG, TMEM199-CDG, ATP6AP1-CDG, MAN1B1-CDG, and PGM1-CDG. Additionally, glycomics is used in clinical laboratory for the characterization of CDGs in patients with an altered sialotransferrin pattern [77].

With all these examples, we can see that with the integration of different "-omics" data, we can improve the diagnosis and the discovery of new diseases and describe new biomarkers as well. However, it still has a high cost-effective price to be implemented in the daily routine in laboratories for the diagnostic of IMD and has mainly been used in the research field. With ongoing technological advancements (both analytical and post-analytical), we hope that in the near future, the number of laboratories with access to this technology will increase.

3.2 Data harmonization

An ongoing challenge in metabolomics is the lack of standardized methods for analyzing untargeted metabolomics data, resulting in varying levels of confidence in metabolite annotation. Metabolite annotation is particularly difficult due to differences in mobile phases, instruments, ionization sources and collision energies, which can lead to variable metabolite fragmentation patterns and product ion intensities [54]. To address this, it is crucial to analyze reference standards under the same experimental conditions as study samples. However, not all necessary standards are commercially available, complicating this process.

Various approaches have been proposed for the analysis and interpretation of metabolomic data, depending on the analytical platform and the specific goals of the analysis. These range from post-analytical normalization procedures within batches to quantify metabolites of interest, particularly relevant for IMD diagnosis and monitoring.

A significant challenge in metabolomics is the presence of batch effects, which can introduce technical variation that must be minimized to enable comparisons between batches. Harmonization of data is necessary to facilitate comparisons across different studies and analytical methods. Different strategies have been proposed to correct for batch effects based on the use of scaling factors, quality control samples, internal standards, and statistical correction strategies [78, 79].

There is a limited availability of reference materials, and laboratories often develop their own reference materials containing metabolites of interest, which can introduce variability. Liu et al. reported on the suitability of three available reference materials (Qstd3, 211 CHEAR, NIST1950) [79].

Data in metabolomics exhibit variability due to heterogeneity in data formats, analysis pipelines, algorithms, and statistical methods used. Standardization efforts are crucial to assess the reliability of these methods and to define standards for reporting metabolomics data effectively [15].

Various publications have addressed quality assessment in untargeted metabolomic studies [80–82]. However, definitive guidelines outlining the minimum quality standards and necessary studies to ensure compliance with these standards, akin to the guidelines proposed by the UK Consortium on Metabolic Phenotyping (MAP/UK) for targeted metabolomics, are still lacking [15]. Additionally, for harmonization between laboratories, especially those focused on the diagnosis and follow-up of IMD, External Quality Control Schemes will be crucial.

4. Lipidomics

Lipidomics profiling, a subset of metabolomics focused on the qualitative and quantitative analysis of lipids, has emerged as a promising new tool for identifying alterations in lipid classes and discovering new biomarker candidates in rare diseases such as IMD [83].

Analytical approaches typically used in lipidomics differ significantly from methods established in metabolomics. Most lipid molecules contain polar/ionic head groups and nonpolar fatty acyl chain(s) (for example, phospholipids and sphingolipids), which results in the formation of amphiphilic molecules with specific physicochemical properties. These properties must be considered during method development including sample preparation, chromatographic separation, and ionization in MS [84].

Lipids play essential roles in wide range of metabolic and physiological processes, and dysregulation of lipid metabolism has been linked to various pathological conditions. Thus, lipids are of interest as diagnostic and predictive disease biomarkers, therapeutic agents, and targets for evaluating treatment response in clinical research [85].

The human lipidome reflects the inter-individual variation in lipid species and has the potential to deepen our understanding of fundamental biological processes, providing broader insights into disease causes and progression. In the clinical setting, lipid measurements have remained unchanged for the past 60 years, typically focusing on total triglycerides, total cholesterol, and its subclasses. This approach results in a poorly described total lipidomic profile, leaving a wealth of information undiscovered. Consequently, the field of clinical lipidomics has become a steadily growing and promising area for addressing congenital and rare diseases [83, 86].

In the past, the lipidomic analysis relied on thin-layer chromatography or gas chromatography after the derivation of polar functionalities. However, the current gold standard is the use of atmospheric pressure ionization MS, either without separation or coupled with liquid-phase separation techniques [84].

Low-resolution platforms rely on triple quadrupole technology to identify lipids by their intact mass and their characteristic fragments, whereas HRMS platforms, primarily using Orbitrap and QTOF technology, additionally provide elemental compositions of parent and fragment masses. The most widely used chromatographic techniques are reversed-phase HPLC and hydrophilic interaction liquid chromatography (HILIC). Reversed-phase chromatography separates lipids by their hydrophobic fatty acid moieties, offering the advantage of separating molecular lipid species within each lipid class and thus increasing identification coverage. In contrast, HILIC separates lipids by their polar headgroups into lipid classes, improving quantitation properties [83].

Current methodology face difficulties in identifying lipid species and patterns due to the overlap of isobaric compounds. Furthermore, when fatty acids are used as precursors, they can be incorporated into complex lipids, elongated, desaturated, or undergo beta-oxidation, resulting in multiple variations of the precursor. To overcome these challenges, novel multidisciplinary approaches, including data science and machine learning techniques, are required. These methods aim to better understand, connect, and interpret different lipid fluxes and elucidate how these lipid fluxes contribute to disease development [86].

Nevertheless, there are few publications of lipidomic studies in IMD. Some examples are the lipidomic study performed by Jaspers et al. on fibroblasts from adrenoleukodystrophy (X-ALD) patients, where they identified new biomarker candidates [87], or the evaluation by Herzog et al. of the use of UPLC-HRMS for detecting patients with different peroxisomal disorders (Zellweger spectrum disorders, rhizomelic chondrodysplasia punctata type 1 and 5, Refsum disease, D-bifunctional protein deficiency and α-methylacyl-CoA racemase deficiency), where they identified novel lipid species for specific peroxisomal diseases purposed as candidate biomarkers and could also see that lipidome was clearly altered in plasma samples from patients with different peroxisomal disorders [88].

5. Conclusions

The improvement in the diagnosis of IMD has been directly related to technological evolution. Specifically, the progress of different chromatography systems coupled with various spectrometers has positively impacted simplifying and shortening analysis times, as well as increasing the sensitivity and specificity of diagnostic tests. We have moved from conducting a single test for one disease to being able to assess different metabolites and diagnose various conditions in a single analysis. Currently, we are facing a new era where hundreds or thousands of metabolites can be analyzed in a sample using HRMS. A paradigm shift in diagnosis is anticipated with the implementation in clinical laboratories in the near future. However, guidelines are needed to ensure the quality assurance of results and reliable quantification, as well as for data harmonization and interpretation. Maybe, the future perspective of metabolomics aims to achieve a single, cost-effective platform that allows simultaneous targeted and non-targeted metabolomic studies, capable of quantifying metabolites of interest for the follow-up of IMD.

The evolution of metabolomics has been aligned with the progress of other omics. Therefore, the development of standardized pipelines for data processing and integration with other '-omic' data would enable a fully comprehensive understanding of each individual patient's physiopathology.

With the progress in genomics, it may become the first-line test for diagnosis, potentially placing metabolomics in a secondary role. This change is expected to occur in newborn screening programs. However, the expansion in genomic studies has led to the identification of rare variants with uncertain significance, requiring biochemical tests to understand the functional impact of these variants. Thus, both omics will be complementary. Perhaps in the future, the metabolome alone will suffice for this purpose, but integrating data from different omics such as metabolomics, transcriptomics, lipidomics, and proteomics will remain essential to predict the pathogenicity of these variants and apply personalized medicine for each patient.

Conflict of interest

The authors declare no conflict of interest.

Appendices and nomenclature

ACMG	American College of Medical Genetics and Genomics
C5DC	glutaryl carnitine
CDG	congenital disorder of glycosylation
CLIA	clinical laboratory improvement amendments
DBS	dried blood spots
EMA	European medicines agency
ERNDIM	European Research Network for evaluation and improvement of screening, Diagnosis, and treatment of Inherited disorders of Metabolism
FDA	food and drug administration
FIA	flow injection analysis
GA-1	glutaric aciduria type I
GC	gas chromatography
GC-MS	gas chromatography- mass spectrometry
HILIC	hydrophilic interaction liquid chromatography
HPLC	high-performance liquid chromatography
HRMS	high-resolution mass spectrometry
IEM	inborn errors of metabolism
IMD	Inherited metabolic disorders
LC	liquid chromatography
MADD	deficiency of medium-chain acyl-CoA dehydrogenase
MS/MS	tandem mass spectrometry
MS	mass spectrometry
NBSP	Newborn screening programs
NGS	next generation sequencing
PKU	phenylketonuria
QTOF	quadrupole-time of flight

UHPLC	ultra-high performance liquid chromatography
VLCADD	deficiency of very long-chain acyl-CoA dehydrogenase
VUS	variants of uncertain clinical significance
WES	whole exome sequencing
X-ALD	X-linked adrenoleukodystrophy

Author details

Clara Oliva Mussarra[1] and Judit García-Villoria[2*]

1 Division of Inborn Errors of Metabolism-IBC, Department of Biochemical and Molecular Genetics, Hospital Clinic of Barcelona, Barcelona, Spain

2 Division of Inborn Errors of Metabolism-IBC, Department of Biochemical and Molecular Genetics, Hospital Clinic of Barcelona, IDIBAPS, CIBERER, Barcelona, Spain

*Address all correspondence to: jugarcia@clinic.cat

References

[1] Fonseca T, Macedo MF. Inherited metabolic disorders: From bench to bedside. Biomedicine. 2024;**12**:174. DOI: 10.3390/biomedicines12010174

[2] AKU SOCIETY. Available from: https://akusociety.org/history-of-aku/ [Accessed: June 21, 2024]

[3] Ferreira CR, Rahman S, Keller M, et al. An international classification of inherited metabolic disorders (ICIMD). Journal of Inherited Metabolic Disease. 2021;**44**:164-177. DOI: 10.1002/jimd.12348

[4] Waters D, Adeloye D, Woolham D, et al. Global birth prevalence and mortality from inborn errors of metabolism: A systematic analysis of the evidence. Journal of Global Health. 2018;**8**:021102. DOI: 10.7189/jogh.08.021102

[5] Saudubray J-M, Mochel F. The phenotype of adult versus pediatric patients with inborn errors of metabolism. Journal of Inherited Metabolic Disease. 2018;**41**:753-756. DOI: 10.1007/s10545-018-0209-9

[6] Lee KN, Uhlmann W, Hipp L, et al. The diagnosis of inborn errors of metabolism in previously undiagnosed adults referred for medical genetics evaluation. Molecular Genetics and Metabolism Reports. 2020;**25**:100653. DOI: 10.1016/j.ymgmr.2020.100653

[7] Saudubray J-M, Baumgartner MR, García-Cazorla A, et al. Inborn Metabolic Diseases: Diagnosis and Treatment. 7th ed. Berlin, Germany: Springer; 2022

[8] Garg U, Smith LD. Biomarkers in Inborn Errors of Metabolism: Clinical Aspects and Laboratory Determination. Cambridge, MA: Academic Press; 2017

[9] Bongaerts M, Bonte R, Demirdas S, et al. Integration of metabolomics with genomics: Metabolic gene prioritization using metabolomics data and genomic variant (CADD) scores. Molecular Genetics and Metabolism. 2022;**136**:199-218. DOI: 10.1016/j.ymgme.2022.05.002

[10] Arnold GL. Inborn errors of metabolism in the 21st century: Past to present. Annals of Translational Medicine. 2018;**6**:467-467. DOI: 10.21037/atm.2018.11.36

[11] Mak J, Cowan TM. Detecting lysosomal storage disorders by glycomic profiling using liquid chromatography mass spectrometry. Molecular Genetics and Metabolism. 2021;**134**:43-52. DOI: 10.1016/j.ymgme.2021.08.006

[12] Tebani A, Abily-Donval L, Afonso C, et al. Clinical metabolomics: The new metabolic window for inborn errors of metabolism investigations in the post-genomic era. International Journal of Molecular Sciences. 2016;**17**:1167. DOI: 10.3390/ijms17071167

[13] Centers for Medicare, Medicaid Services. Clinical Laboratory Improvement Amendments of 1988 (CLIA). 1988. Available from: https://www.cms.gov/Regulations-and-Guidance/Legislation/CLIA [Accessed: July 25, 2024]

[14] ERNDIM HISORY. Available from: https://www.erndim.org/about-us/history-of-erndim/ [Accessed: May 30, 2024]

[15] Sarmad S, Viant MR, Dunn WB, et al. A proposed framework to evaluate

the quality and reliability of targeted metabolomics assays from the UK Consortium on Metabolic Phenotyping (MAP/UK). Nature Protocols. 2023;**18**:1017-1027. DOI: 10.1038/s41596-022-00801-8

[16] Hommes FA. Techniques in Diagnostic Human Biochemical Genetics. New York: Wiley-Liss; 1991

[17] Beckman AO, Gallaway WS, Kaye W, et al. History of spectrophotometry at Beckman instruments, Inc. Analytical Chemistry. 1977;**49**:280A-300A

[18] Guthrie R, Susi A. A simple phenylalanine method for detecting phenylketonuria in large populations of newborn infants. Pediatrics. 1963;**32**:338-343

[19] la Marca G, Carling Rachel S, Moat Stuart J, et al. Current state and innovations in newborn screening: Continuing to do good and avoid harm. International Journal of Neonatal Screen. 2023;**9**:15. DOI: 10.3390/ijns9010015

[20] Rahman M, Ramaswamy M. Metabolomics used in clinical diagnosis related to newborn screening. In: Metabolomics. Amsterdam, Netherlands: Elsevier; 2023. pp. 121-171

[21] Stanford M, Spackman DH, Stein WH. Chromatography of amino acids on sulfonated polystyrene resins. An improved system. Analytical Chemistry. 1958;**30**:1185-1190

[22] Bieber LL, Kerner J. Short-chain acylcarnitines: Identification and quantitation. Methods in Enzymology. 1986;**123**:264-276

[23] Goodman SI, Markey SP. Diagnosis of organic acidemias by gas chromatography--Mass spectrometry. Laboratory and Research Methods in Biology and Medicine. 1981;**6**:1-158

[24] Thompson JA, Markey SP. Quantitative metabolic profiling of urinary organic acids by gas chromatography-mass spectrometry. Comparison of isolation methods. Analytical Chemistry. 1975;**47**:1313-1321

[25] Hommes F, Kuipers J, Elema J, et al. Propionic acidemia, a new inborn error of metabolism. Pediatric Research. 1968;**2**:519-524. DOI: 10.1203/00006450-196811000-00010

[26] Stokke O, Eldjarn L, Norum K, Steen-Johnsen J, Halvorsen S. Methylmalonic aciduria: A new inborn error of metabolism which may cause fatal acidosis in the neonatal period. Scandinavian Journal of Clinical and Laboratory Investigation. 1967;**20**:313

[27] Oberholzer VG, Levin B, Burgess EA, et al. Methylmalonic aciduria. An inborn error of metabolism leading to chronic metabolic acidosis. Archives of Disease in Childhood. 1967;**42**:492-504. DOI: 10.1136/adc.42.225.492

[28] Tanaka K, Budd MA, Efron ML, et al. Isovaleric acidemia: A new genetic defect of leucine metabolism. Proceedings of the National Academy of Sciences. 1966;**56**:236-242. DOI: 10.1073/pnas.56.1.236

[29] Dancis J, Levitz M, Miller SB, et al. Maple syrup urine disease. British Medical Journal. 1959;**1**:91-93. DOI: 10.1136/bmj.1.5114.91

[30] Przyrembel H, Wendel U, Becker K, et al. Glutaric aciduria type II: Report on a previously undescribed metabolic disorder. Clinica Chimica Acta. 1976;**66**:227-239. DOI: 10.1016/0009-8981(76)90060-7

[31] Millington DS, Kodo N, Norwood DL, et al. Tandem mass spectrometry: A new method for acylcarnitine profiling with potential for neonatal screening for inborn errors of metabolism. Journal of Inherited Metabolic Disease. 1990;**13**:321-324. DOI: 10.1007/BF01799385

[32] Nicholson JK, Lindon JC, Holmes E. 'Metabonomics': Understanding the metabolic responses of living systems to pathophysiological stimuli via multivariate statistical analysis of biological NMR spectroscopic data. Xenobiotica. 1999;**29**:1181-1189. DOI: 10.1080/004982599238047

[33] Carling RS, Whyte E, John C, et al. Improving harmonization and standardization of expanded newborn screening results by optimization of the legacy flow injection analysis tandem mass spectrometry methods and application of a standardized calibration approach. Clinical Chemistry. 2022;**68**:1075-1083. DOI: 10.1093/clinchem/hvac070

[34] Moldoveanu C, David V. Progress in technology of the chromatographic columns in HPLC. In: Analytical Liquid Chromatography—New Perspectives. London, UK: IntechOpen; 2022. Epub ahead of print 25 August 2022. DOI: 10.5772/intechopen.104123

[35] Zotou A. An overview of recent advances in HPLC instrumentation. Open Chemistry. 2012;**10**:554-569

[36] Hertzog A, Selvanathan A, Devanapalli B, et al. A narrative review of metabolomics in the era of "-omics": Integration into clinical practice for inborn errors of metabolism. Translational Pediatrics. 2022;**11**:1704-1716. DOI: 10.21037/tp-22-105

[37] Rebollido-Fernandez MM, Castiñeiras DE, Dolores Bóveda M, et al. Development of electrospray ionization tandem mass spectrometry methods for the study of a high number of urine markers of inborn errors of metabolism. Rapid Communications in Mass Spectrometry. 2012;**26**:2131-2144. DOI: 10.1002/rcm.6325

[38] Pajares-García S, González de Aledo-Castillo JM, Flores-Jiménez JE, et al. Analysis of a second-tier test panel in dried blood spot samples using liquid chromatography-tandem mass spectrometry in Catalonia's newborn screening programme. Clinical Chemistry and Laboratory Medicine (CCLM). 2024;**62**:493-505. DOI: 10.1515/cclm-2023-0216

[39] Coene KLM, Kluijtmans LAJ, van der Heeft E, et al. Next-generation metabolic screening: Targeted and untargeted metabolomics for the diagnosis of inborn errors of metabolism in individual patients. Journal of Inherited Metabolic Disease. 2018;**41**:337-353. DOI: 10.1007/s10545-017-0131-6

[40] Vinayavekhin N, Saghatelian A. Untargeted metabolomics. Current Protocols in Molecular Biology. 2010;**90**:30.1.1-30.1.24. Epub ahead of print April 2010. DOI: 10.1002/0471142727.mb3001s90

[41] van Scherpenzeel M, Steenbergen G, Morava E, et al. High-resolution mass spectrometry glycoprofiling of intact transferrin for diagnosis and subtype identification in the congenital disorders of glycosylation. Translational Research. 2015;**166**:639-649.e1. DOI: 10.1016/j.trsl.2015.07.005

[42] Guillard M, Morava E, van Delft FL, et al. Plasma N-glycan profiling by mass spectrometry for congenital disorders of glycosylation type II. Clinical Chemistry. 2011;**57**:593-602. DOI: 10.1373/clinchem.2010.153635

[43] Roberts LD, Souza AL, Gerszten RE, et al. Targeted metabolomics. Current Protocols in Molecular Biology. 2012;**98**:30.2.1-30.2.24. Epub ahead of print 15 April 2012. DOI: 10.1002/0471142727.mb3002s98

[44] Dunn WB, Broadhurst DI, Atherton HJ, et al. Systems level studies of mammalian metabolomes: The roles of mass spectrometry and nuclear magnetic resonance spectroscopy. Chemical Society Reviews. 2011;**40**:387-426. DOI: 10.1039/b906712b

[45] Wishart DS. Emerging applications of metabolomics in drug discovery and precision medicine. Nature Reviews. Drug Discovery. 2016;**15**:473-484. DOI: 10.1038/nrd.2016.32

[46] Dudley E, Yousef M, Wang Y, et al. Targeted metabolomics and mass spectrometry. Advances in Experimental Medicine and Biology. 2010;**678**:45-83. DOI: 10.1016/B978-0-12-381264-3.00002-3

[47] Griffin JL, Nicholls AW. Metabolomics as a functional genomic tool for understanding lipid dysfunction in diabetes, obesity and related disorders. Pharmacogenomics. 2006;**7**:1095-1107. DOI: 10.2217/14622416.7.7.1095

[48] Hoegen B, Zammit A, Gerritsen A, et al. Metabolomics-based screening of inborn errors of metabolism: Enhancing clinical application with a robust computational pipeline. Metabolites. 2021;**11**:568

[49] Liu N, Xiao J, Gijavanekar C, et al. Comparison of untargeted metabolomic profiling vs traditional metabolic screening to identify inborn errors of metabolism. JAMA Network Open. 2021;**4**:e2114155. DOI: 10.1001/jamanetworkopen.2021.14155

[50] Kennedy AD, Miller MJ, Beebe K, et al. Metabolomic profiling of human urine as a screen for multiple inborn errors of metabolism. Genetic Testing and Molecular Biomarkers. 2016;**20**:485-495. DOI: 10.1089/gtmb.2015.0291

[51] Miller MJ, Kennedy AD, Eckhart AD, et al. Untargeted metabolomic analysis for the clinical screening of inborn errors of metabolism. Journal of Inherited Metabolic Disease. 2015;**38**:1029-1039. DOI: 10.1007/s10545-015-9843-7

[52] Kennedy AD, Wittmann BM, Evans AM, et al. Metabolomics in the clinic: A review of the shared and unique features of untargeted metabolomics for clinical research and clinical testing. Journal of Mass Spectrometry. 2018;**53**:1143-1154. DOI: 10.1007/s10545-015-9843-7

[53] Evans AM, DeHaven CD, Barrett T, et al. Integrated, nontargeted ultrahigh performance liquid chromatography/electrospray ionization tandem mass spectrometry platform for the identification and relative quantification of the small-molecule complement of biological systems. Analytical Chemistry. 2009;**81**:6656-6667. DOI: 10.1021/ac901536h

[54] Schrimpe-Rutledge AC, Codreanu SG, Sherrod SD, et al. Untargeted metabolomics strategies—Challenges and emerging directions. Journal of the American Society for Mass Spectrometry. 2016;**27**:1897-1905. DOI: 10.1007/s13361-016-1469-y

[55] Becker S, Kortz L, Helmschrodt C, et al. LC–MS-based metabolomics in the clinical laboratory. Journal of Chromatography B. 2012;**883-884**:68-75. DOI: 10.1016/j.jchromb.2011.10.018

[56] Alseekh S, Aharoni A, Brotman Y, et al. Mass spectrometry-based

metabolomics: A guide for annotation, quantification and best reporting practices. Nature Methods. 2021;**18**:747-756. DOI: 10.1038/s41592-021-01197-1

[57] Sumner LW, Amberg A, Barrett D, et al. Proposed minimum reporting standards for chemical analysis. Metabolomics. 2007;**3**:211-221. DOI: 10.1007/s11306-007-0082-2

[58] Perez de Souza L, Alseekh S, Scossa F, et al. Ultra-high-performance liquid chromatography high-resolution mass spectrometry variants for metabolomics research. Nature Methods. 2021;**18**:733-746. DOI: 10.1038/s41592-021-01116-4

[59] Wurth R, Turgeon C, Stander Z, et al. An evaluation of untargeted metabolomics methods to characterize inborn errors of metabolism. Molecular Genetics and Metabolism. 2024;**141**:108115. DOI: 10.1016/j.ymgme.2023.108115

[60] Steinbusch LKM, Wang P, Waterval HWAH, et al. Targeted urine metabolomics with a graphical reporting tool for rapid diagnosis of inborn errors of metabolism. Journal of Inherited Metabolic Disease. 2021;**44**:1113-1123. DOI: 10.1002/jimd.12385

[61] Sebaa R, AlMogren M, Alseraty W, et al. Untargeted metabolomics identifies biomarkers for MCADD neonates in dried blood spots. International Journal of Molecular Sciences. 2023;**24**:9657. DOI: 10.3390/ijms24119657

[62] Sebaa R, AlMalki RH, Alseraty W, et al. A distinctive metabolomics profile and potential biomarkers for very long acylcarnitine dehydrogenase deficiency (VLCADD) diagnosis in newborns. Metabolites. 2023;**13**:725. DOI: 10.3390/metabo13060725

[63] Jacob M, Malkawi A, Albast N, et al. A targeted metabolomics approach for clinical diagnosis of inborn errors of metabolism. Analytica Chimica Acta. 2018;**1025**:141-153. DOI: 10.1016/j.aca.2018.03.058

[64] Oliver S. Systematic functional analysis of the yeast genome. Trends in Biotechnology. 1998;**16**:373-378. DOI: 10.1016/s0167-7799(98)01214-1

[65] Goodacre R, Vaidyanathan S, Dunn WB, et al. Metabolomics by numbers: Acquiring and understanding global metabolite data. Trends in Biotechnology. 2004;**22**:245-252. DOI: 10.1016/j.tibtech.2004.03.007

[66] Graham E, Lee J, Price M, et al. Integration of genomics and metabolomics for prioritization of rare disease variants: A 2018 literature review. Journal of Inherited Metabolic Disease. 2018;**41**:435-445. DOI: 10.1007/s10545-018-0139-6

[67] Houten SM, Denis S, te Brinke H, et al. Mitochondrial NADP(H) deficiency due to a mutation in NADK2 causes dienoyl-CoA reductase deficiency with hyperlysinemia. Human Molecular Genetics. 2014;**23**:5009-5016. DOI: 10.1093/hmg/ddu218

[68] Kerkhofs MHPM, Haijes HA, Willemsen AM, et al. Cross-omics: Integrating genomics with metabolomics in clinical diagnostics. Metabolites. 2020;**10**:206. DOI: 10.3390/metabo10050206

[69] Alaimo JT, Glinton KE, Liu N, et al. Integrated analysis of metabolomic profiling and exome data supplements sequence variant interpretation, classification, and diagnosis. Genetics in Medicine. 2020;**22**:1560-1566. DOI: 10.1038/s41436-020-0827-0

[70] Graham Linck EJ, Richmond PA, Tarailo-Graovac M, et al. metPropagate: Network-guided propagation of metabolomic information for prioritization of metabolic disease genes. NPJ Genomic Medicine. 2020;**5**:25. DOI: 10.1038/s41525-020-0132-5

[71] Kennedy AD, Pappan KL, Donti T, et al. 2-pyrrolidinone and succinimide as clinical screening biomarkers for GABA-transaminase deficiency: Anti-seizure medications impact accurate diagnosis. Frontiers in Neuroscience. 2019;**13**:394. Epub ahead of print 8 May 2019. DOI: 10.3389/fnins.2019.00394

[72] Tang C, Li L, Chen T, et al. Newborn screening for inborn errors of metabolism by next-generation sequencing combined with tandem mass spectrometry. International Journal of Neonatal Screen. 2024;**10**:28. DOI: 10.3390/ijns10020028

[73] Brnich SE, Abou Tayoun AN, Couch FJ, et al. Recommendations for application of the functional evidence PS3/BS3 criterion using the ACMG/AMP sequence variant interpretation framework. Genome Medicine. 2020;**12**:3. DOI: 10.1186/s13073-019-0690-2

[74] Richards S, Aziz N, Bale S, et al. Standards and guidelines for the interpretation of sequence variants: A joint consensus recommendation of the American College of Medical Genetics and Genomics and the Association for Molecular Pathology. Genetics in Medicine. 2015;**17**:405-424. DOI: 10.1038/gim.2015.30

[75] Van Scherpenzeel M, Willems E, Lefeber DJ. Clinical diagnostics and therapy monitoring in the congenital disorders of glycosylation. Glycoconjugate Journal. 2016;**33**:345-358. DOI: 10.1007/s10719-015-9639-x

[76] Argov Z, Caraco Y, Lau H, et al. Aceneuramic acid extended release administration maintains upper limb muscle strength in a 48-week study of subjects with GNE myopathy: Results from a phase 2, randomized, controlled study. Journal of Neuromuscular Diseases. 2016;**3**:49-66. DOI: 10.3233/JND-159900

[77] Abu Bakar N, Lefeber DJ, van Scherpenzeel M. Clinical glycomics for the diagnosis of congenital disorders of glycosylation. Journal of Inherited Metabolic Disease. 2018;**41**:499-513. DOI: 10.1007/s10545-018-0144-9

[78] Rampler E, Hermann G, Grabmann G, et al. Benchmarking non-targeted metabolomics using yeast-derived libraries. Metabolites. 2021;**11**:160. DOI: 10.3390/metabo11030160

[79] Liu KH, Nellis M, Uppal K, et al. Reference standardization for quantification and harmonization of large-scale metabolomics. Analytical Chemistry. 2020;**92**:8836-8844. DOI: 10.1021/acs.analchem.0c00338

[80] Dunn WB, Broadhurst DI, Edison A, et al. Quality assurance and quality control processes: Summary of a metabolomics community questionnaire. Metabolomics. 2017;**13**:50. DOI: 10.1007/s11306-017-1188-9

[81] Beger RD, Dunn WB, Bandukwala A, et al. Towards quality assurance and quality control in untargeted metabolomics studies. Metabolomics. 2019;**15**:4. DOI: 10.1007/s11306-018-1460-7

[82] Broadhurst D, Goodacre R, Reinke SN, et al. Guidelines and considerations for the use of system suitability and quality control samples in mass spectrometry assays applied in

untargeted clinical metabolomic studies. Metabolomics. 2018;**14**:72. DOI: 10.1007/s11306-018-1367-3

[83] Zandl-Lang M, Plecko B, Köfeler H. Lipidomics—Paving the road towards better insight and precision medicine in rare metabolic diseases. International Journal of Molecular Sciences. 2023;**24**:1709. DOI: 10.3390/ijms24021709

[84] Holčapek M, Liebisch G, Ekroos K. Lipidomic analysis. Analytical Chemistry. 2018;**90**:4249-4257. DOI: 10.1021/acs.analchem.7b05395

[85] Parker BL, Calkin AC, Seldin MM, et al. An integrative systems genetic analysis of mammalian lipid metabolism. Nature. 2019;**567**:187-193. DOI: 10.1038/s41586-019-0984-y

[86] Zandl-Lang M. Tracing the lipidome in inborn errors of metabolism. Biochimica et Biophysica Acta (BBA)—Molecular and Cell Biology of Lipids. 2024;**1869**:159491. DOI: 10.1016/j.bbalip.2024.159491

[87] Jaspers YRJ, Meyer SW, Pras-Raves ML, et al. Four-dimensional lipidomics profiling in X-linked adrenoleukodystrophy using trapped ion mobility mass spectrometry. Journal of Lipid Research. 2024;**65**:100567. DOI: 10.1016/j.jlr.2024.100567

[88] Herzog K, Pras-Raves ML, Ferdinandusse S, et al. Plasma lipidomics as a diagnostic tool for peroxisomal disorders. Journal of Inherited Metabolic Disease. 2018;**41**:489-498. DOI: 10.1007/s10545-017-0114-7

Chapter 8

Advances in Applications of High-Performance Liquid Chromatography in the Analysis of Herbal Products

Ali Gamal Al-Kaf, Rana Abdullah Al-Robaidi and Hend Ahmed Al-Haj

Abstract

High-performance liquid chromatography (HPLC) is widely regarded as one of the most effective and adaptable methods for separating and analyzing the components of herbal mixtures. It is utilized to identify and quantify phyto-components within these mixtures. This chapter focuses on the utilization of HPLC techniques, including ultra-performance liquid chromatography (UPLC) or ultra-high-performance liquid chromatography (UHPLC), in the analysis of various herbal products. It provides several specific examples of protocols for such analysis and offers a brief overview of available HPLC techniques and methods. Additionally, this chapter includes step-by-step protocols for the chemical profiling or fingerprinting of herbs, herbal mixtures, and herbal products.

Keywords: HPLC, herbal mixtures, phyto-components, UPLC, UHPLC, herbal products

1. Introduction

Herbal remedies, widely used in healthcare worldwide, contain complex combinations of chemicals sourced from plants. However, their effectiveness is often hindered by poor oral absorption. The World Health Organization (WHO) estimates that approximately 80% of the global population continues to rely on herbs and other traditional medicines to address their primary healthcare requirements. Herbal formulations have gained widespread acceptance as therapeutic remedies for various conditions, including diabetes, arthritis, liver diseases, cough relief, memory enhancement, and adaptogens.

According to the World Health Organization's (WHO) definition, herbal medicines can be categorized into three main groups: raw plant material, processed plant material, and medicinal herbal products. Herbal drugs refer to finished and labeled products containing active ingredients derived from plants, including aerial or underground plant parts or other plant materials, whether in their natural state or as plant preparations. The utilization of herbal medicines has seen a significant rise, aligning with the global inclination toward natural therapies. Individuals consume herbal medicine products, categorized as dietary supplements, with the aim of enhancing their health. These products are available in various forms, such as tablets, capsules, powders, teas, extracts, and even fresh or dried plants. Traditional beliefs often regard herbal remedies as safe, which has led to their increased usage by individuals without medical prescriptions. However, certain herbal products may lead to health complications, prove ineffective, and occasionally even interact negatively with other medications. Establishing standardization in herbal formulations is vital for evaluating drug quality because it enables the measurement of concentrations of active principles within the products. Assessing the quality of herbal preparations is a crucial necessity for industries and other entities involved in the production and distribution of Ayurvedic and herbal products [1]. Herbal drug technology focuses on transforming botanical materials into medicines. It highlights the significance of standardization and quality control, accomplished by effectively blending modern scientific techniques with traditional knowledge. A secure method known as reverse pharmacology, which draws on traditional knowledge databases, can be employed to develop scientifically validated and technologically standardized herbal medicines. Certainly, incorporating reverse pharmacology into the process of developing scientifically validated and technologically standardized herbal medicines holds great potential to advance drug discovery, development, and therapeutic outcomes. According to regulatory guidelines and pharmacopeias, ensuring the quality control and standardization of botanical materials involves macroscopic and microscopic evaluations, along with chemical profiling. Thin layer chromatography (TLC) and high-performance thin layer chromatography (HPTLC) serve as valuable methods for qualitatively assessing small quantities of impurities. Additionally, a range of analytical techniques, including volumetric analysis, gravimetric determinations, gas chromatography (GC), column chromatography (CC), high-performance liquid chromatography (HPLC), and spectrophotometric methods, are frequently employed for quality control and standardization purposes [2]. Therefore, it is essential for each individual herb to undergo quality checks to ensure compliance with quality standards and consistent delivery of desired properties. Standardization ensures that products are dependable in terms of quality, effectiveness, performance, and safety [3].

2. Chromatography and chemical fingerprints of herbal medicines

Chromatography can be described as a method for separating and identifying elements, compounds, or mixtures into their individual components. This is achieved through the utilization of a stationary phase and a mobile phase. Adhering to stringent guidelines is imperative to ensure the successful manufacture of high-quality herbal medication. Included in these requirements are accurate botanical identification, screening for phytochemicals, and establishing standardization procedures. Quality assurance and standardization processes for herbal medicines encompass multiple stages. Ensuring the quality control of herbal medicines unquestionably relies on the source and quality of raw materials, along with the adoption of effective agricultural methods and manufacturing processes. They play essential roles in ensuring the quality and consistency of herbal preparations. The chemical makeup of herbal products can vary depending on factors like when they are harvested, which parts of the plant are collected, the time of year, where the plants are from, how they are dried, and other factors that can affect them. Therefore, it becomes essential to identify the majority of phytochemical constituents present in herbal products to ensure the credibility and consistency of pharmacological and clinical studies. This assists in understanding their bioactive characteristics and the possible side effects of active ingredients, thus enabling better quality control of the products. The idea of phytoequivalence originated in Germany as a means to guarantee the consistency and uniformity of herbal products. According to this principle, an herbal product should possess a chemical profile resembling a chromatographic fingerprint, which can be compared to the profile of a clinically proven reference product. Essentially, a chromatographic fingerprint of an herbal remedy is a pattern that shows the typical chemical components of the extract, which have pharmacologically active and/or distinctive chemical traits. The chromatographic profile should demonstrate both "integrity" and "fuzziness" or "consistency" and "variations," accurately representing the herbal drug being analyzed. Through chromatographic fingerprints, herbal medicines can be authenticated and identified precisely ("integrity"), even when the quantity and/or concentration of distinct chemical constituents vary among different drug samples (thus, "fuzziness"). Additionally, these chromatographic fingerprints could effectively illustrate both the "similarity" and "variations" among different samples. Hence, it is imperative to comprehensively analyze multiple constituents within herbal drug extracts rather than focusing solely on one or two marker components when assessing the quality of herbal products. Nevertheless, herbal drugs and their extracts typically contain numerous unidentified components, many of which are present in low concentrations. Furthermore, variability within the same herbal materials is commonly observed. As a result, obtaining dependable chromatographic fingerprints that accurately represent pharmacologically active and chemically distinctive components is a challenging and non-trivial task. Thankfully, chromatography provides a highly effective separation capability, enabling the intricate chemical constituents in herbal extracts to be divided into numerous relatively simple sub-fractions [4]. The majority of active ingredients in medicines have drawn inspiration from natural products (NPs). This success in drug discovery can likely be attributed to several factors, including the high chemical diversity of natural products, the impact of evolutionary pressures in generating biologically active molecules, and the structural similarity of protein targets across various species. The vast chemical diversity of natural products is intricately linked with a wide variability in their inherent physicochemical properties, possessing a significant challenge to their universal detection [5]. Over the years, HPLC has undergone significant advancements in terms of convenience, speed, availability of column stationary phases, sensitivity, versatility across a wide range of sample matrices, and the capability to integrate chromatographic methods with spectroscopic detectors [6]. The evolution of columns with varied phase chemistry, particularly reversed phase, from a chromatographic perspective has facilitated the separation of virtually all types of natural products. Recent advancements in HPLC, such as the introduction of highly pH-stable phases, sub-2-μm particles, and monolith columns, have significantly enhanced the performance of HPLC systems in terms of resolution, speed, and reproducibility [7]. Through the coupling of columns, it is possible to achieve efficiencies surpassing 100,000 plates and peak capacities exceeding 900 [8]. To isolate crude extracts, you can introduce either unrefined mixtures or samples concentrated through basic solid-phase extraction (SPE) or liquid-liquid extraction (LLE) into the chromatographic setup. Most separations are carried out using reversed-phase chromatography with C18 material, employing either the ACN-H_2O (acetonitrile in water) or MeOH-H_2O (methanol in water) solvent system in gradient elution mode. To improve separation effectiveness, different modifiers are incorporated into the mobile phase, potentially influencing detection sensitivity to a significant degree. In systems incorporating multiple detectors connected with hyphens (multi-hyphenated systems), it's essential to have an eluent composition that performs effectively across all detectors [9]. High-performance liquid chromatography (HPLC) is widely used for natural product profiling and fingerprinting, quantitative analysis, and ensuring quality control. Here's a concise overview of the available methods for detecting natural products in this context. We'll delve into each detector, examining their applicable range of natural products, sensitivity, selectivity, and potential for offering online structural information. There are two primary types of detectors: the first category includes basic ones utilized to capture chromatographic traces for profiling or quantification, such as UV, ELSD, and ECD. The second category consists of detectors used in hyphenated systems, which generate multidimensional data by integrating chromatographic and spectroscopic information for online identification and dereplication, examples of which are UV-DAD, MS, and NMR [10].

3. Detection methods for profiling and quantification

3.1 HPLC-UV (ultraviolet detection)

UV detection stands out as the simplest and most commonly employed among all HPLC detectors. Its sensitivity depends on two main factors: the magnitude of the extinction coefficient of the analyte at a particular wavelength and the path length of the light passing through the UV cell. As the path length increases, sensitivity typically improves, but there's a need to strike a balance with the cell volume to prevent peak dispersion [11]. Although HPLC-UV has its limitations, particularly when dealing with natural products that do not have UV chromophores, it still stands out as the ideal balance of sensitivity, linearity, versatility, and reliability sought after in all LC detectors developed thus far. The majority of natural products absorb UV light within the range of 200–550 nm, which includes all substances with one or more double bonds and those with unshared electrons. Therefore, even compounds with weak chromophores, such as triterpene glycosides, can be efficiently detected by UV at shorter wavelengths (for example, 203 nm) [12]. In such situations, it's vital to steer clear of mobile-phase components with high UV cut-offs, as they might obscure the detection of natural products with weak chromophores [13]. Three types of UV detectors exist: fixed wavelength, multiple wavelengths, and photodiode array (DAD). The fixed-wavelength detector, the most economical option, offers higher intrinsic sensitivity by emitting light at specific wavelengths using pre-set lamps. The detection limits (LODs) can reach 10^{-8} g/mL, with a linear dynamic range covering approximately three orders of magnitude. Often, the separation of raw extracts requires gradient elution, which could potentially induce a baseline shift at lower wavelengths. To tackle this issue, one can utilize modifiers with low UV cut-offs, such as phosphate buffers or trifluoroacetic acid (TFA). However, it's important to note that these modifiers are unsuitable for multi-hyphenated systems like HPLC-MS because of their non-volatility and the risk of ion suppression, so it's advisable to steer clear of them in such setups [14].

3.2 HPLC-FD (fluorescence detection)

Fluorescence detection (FD) offers significantly improved sensitivity and selectivity compared to HPLC-UV. In fluorescence, a molecule absorbs a photon and then emits another photon with a longer wavelength. The difference in wavelengths between absorption and emission contributes to increased selectivity, while measuring fluorescent light against a minimal background enhances the signal-to-noise ratio (S/N ratio). However, it's important to note that only a restricted range of natural substances emit fluorescence within a practical spectrum of wavelengths. Nevertheless, many substances can be induced to fluoresce by creating suitable derivatives [11]. There has been a recent publication detailing a method for detecting aflatoxins in herbs using fluorescence, with a focus on quantification [15]. By integrating sample purification through immunoaffinity columns with HPLC-FD, it becomes feasible to detect aflatoxin levels ranging from 7 to 20 μg/kg within the plant material. Moreover, HPLC-FD has been employed to screen the primary benzophenanthridine alkaloids produced by cell cultures of *Eschscholtzia californica* [16].

3.3 HPLC-CL (chemiluminescence)

Chemiluminescence is similar to fluorescence in that it is the non-thermal emission of light from an electrically stimulated molecule or atom resulting from a chemical process at room temperature. Chemiluminescent (CL) nitrogen detection is a relatively recent method that offers exceptionally high sensitivity down to the femtogram level for detecting nitrogen-containing compounds, including a broad spectrum of medications. Many chemicals can only be detected by chemiluminescence after undergoing derivatization since they are not chemiluminescent by nature. The use of chemiluminescence detection has increased due to the discovery of chemiluminogenic chemicals, especially in pharmaceutical and biomedical settings [17]. The chemiluminogenic reagents are usually added after the column in HPLC applications. As an illustration, the addition of potassium hexacyanoferrate (III) makes it possible to identify different carboxylic acids by looking for N-(4-aminobutyl)-Derivatives of N-ethylisoluminol [18]. Despite its high sensitivity, the method has seldom been employed for analyzing crude plant extracts. However, as indicated by a study, HPLC-CL effectively identified flavonoids with a sensitivity of 3 ng/mL in phytopharmaceuticals such as *Hippophae rhamnoides*. The approach relied on the flavonols within the cerium (IV)-rhodamine 6G system to amplify chemiluminescence in a sulfuric acid medium [19].

3.4 HPLC-ECD (electrochemical detection)

Many natural products contain electroactive groups, making them easily measurable and detectable by LC with electrochemical detection (ECD). This detection method is cost-effective, sensitive, selective, and widely adopted. While it may not provide the same level of selectivity as fluorescence detection, it can be used for a diverse array of analytes in either oxidation or reduction modes, making it suitable for a broad range of compounds. Numerous functional groups demonstrate sensitivity to oxidation, such as phenols, aromatics, amines, thiols, and quinolines, while several are responsive to reduction, including olefins, esters, ketones, aldehydes, ethers, and quinones. Electrochemical detection (ECD) distinguishes itself from other detection methods by altering the sample during analysis. In HPLC-ECD configurations, cells typically incorporate three electrodes: the working electrode, counter electrode, and reference electrode. While these electrodes can be arranged in various geometries, two main designs are commonly used in coulometric systems: in one, the eluent passes through the electrode, while in another, the eluent flows past it in amperometric systems. In ECD detection, the typical approach involves maintaining the potential of the working electrode constant relative to the potential of the electrolyte, which is monitored by the reference electrode. The constant potential difference applied between the working and reference electrodes facilitates the electrochemical reaction, and the resulting current is measured as a function of elution time. As a result, detection limits at the picomole level can be achieved. The selectivity of ECD depends on factors such as the available potential range, the concentration of active compounds within this range, and the half-widths of the individual signals [13]. Pulsed techniques employ multi-step potential-time waveforms to achieve amperometric/coulometric detection, ensuring consistent and reproducible electrode activity. These methods are also utilized for detection purposes [20].

3.5 HPLC-RID (refractive index detection)

HPLC-RID was used to evaluate the sugar levels in various dry extracts of *Hypericum perforatum* after thorough solid-phase extraction (SPE) purification [21]. In recent advancements, the terpene trilactones found in *Ginkgo biloba* were accurately quantified using HPLC-RID, employing a meticulously crafted sample preparation technique focused on the ionization of these terpenes. This method yielded outstanding results with a relative standard deviation (RSD) of less than 3%, a limit of detection (LOD) of 0.1 μg, and a linear range spanning from 0.1 to 12 μg [22].

3.6 HPLC-ELSD (evaporative light-scattering detection)

ELSD is indeed a quasi-universal detector for liquid chromatography, capable of detecting any analyte less volatile than the mobile phase, irrespective of the optical, electrochemical, or other properties of the analyte [23]. In this detection system, the eluent is atomized using a nitrogen flow, and the resulting mist is directed through a heated drift tube where volatile components and solvents are vaporized. The remaining solid residue is then introduced into a detection cell, where a beam of light is projected onto the particles. This interaction causes the scattering of the incoming light, which is subsequently captured by either a photodiode or a photomultiplier for detection. The nebulizer gas flow rate and the drift tube temperature are the most crucial parameters influencing the ELSD signal response [24]. The gas flow rate affects the size of droplets in the effluent from the column before evaporation. Higher flow rates result in the formation of smaller aerosol droplets, which decrease light scattering and consequently reduce sensitivity. Nevertheless, higher gas flow rates also contribute to a more consistent baseline. The temperature of the drift tube assists in evaporating the nebulized aerosol, facilitating the selective measurement of the light-scattering phenomenon of the non-volatile solute. Elevated temperatures are necessary for mobile phases containing a high proportion of aqueous content or non-volatile solutes compared to semi-volatile compounds. Indeed, the latest generation of ELSD can vaporize the eluent at low temperatures, making it easier to detect semi-volatile analytes. It's crucial to optimize these parameters to ensure the optimal signal-to-noise ratio (S/N) is achieved [25]. In applications related to natural products (NPs), ELSD has been primarily utilized for the detection of compounds with weak chromophores, such as terpenes, in both aglycone and glycosidic forms [26], saponins [27], and some alkaloids [28]. The method has also been applied in conjunction with HPLC-MS for screening natural product libraries [29]. In this scenario, a diverse array of products, including UV-active phenolics, were scrutinized, and an ELSD signal was detected for all compounds with more than 100 nanograms injected. ELSD has been utilized for quantitatively determining the pertinent active components of herbal medications. A novel technique for detecting 14 ginsenosides in Korean red ginseng, employing digoxin as an internal standard, has been validated and applied to various ginseng products [30].

3.7 HPLC-CAD (charged aerosol detection)

Dixon and Peterson introduced charged aerosol detection (CAD), an inventive method for universal HPLC detection that was developed from ELSD, in 2002 [31]. The early phases of the CAD working concept are similar to those of ELSD. In contrast to ELSD, however, CAD uses a corona discharge needle to charge the dry particle stream, and an electrometer is used to measure the resulting electrical charge. UV detection usually outperforms CAD or ELSD for detecting natural products with strong chromophores. However, these alternative detection methods are better suited for analyzing non-UV-absorbing compounds, and their performance is notably superior in such instances [32]. To the best of our knowledge, natural product detection in crude extracts has not made use of CAD, a relatively new detection approach. On the other hand, it has been widely used for the analysis of different pharmaceutical goods. Only two substances with relatively low molecular weights escaped detection by both CAD and ELSD during the examination of a library containing over 700 compounds within the 150–700 Da range [33]. It's noteworthy that CAD has been used to successfully analyze triacylglycerols (TGs) in plant oils quantitatively [30].

3.8 HPLC-FID (flame ionization detection)

The capacity of flame ionization detection (FID) to identify all carbon-containing compounds makes it a common detector for GC and should be mentioned while discussing universal detection methods for HPLC.

Fascinatingly, FID has also been modified for application in HPLC. However, the mobile phases in LC contain organic compounds that may obstruct FID detection in both normal and reversed-phase modes [34]. Utilizing superheated water in HTLC mode is an additional choice. Liquid water's polarity drastically reduces under pressure and at temperatures between 80°C and 250°C, making it suitable to substitute the organic modifier and function as the only eluent [35]. HTLC-FID has been used extremely sparingly, despite reports of successful uses, mostly for the separation of alcohols, carbohydrates, and amino acids [36].

3.9 HPLC-MS (mass spectrometry)

The utilization of HPLC-MS for the examination and characterization of natural products is widely acknowledged as a significant advancement [37]. Employing a mass spectrometer as an HPLC detector provides outstanding sensitivity and selectivity for analyzing natural products (NPs) in complex biological samples, despite its significant expense. Additionally, MS detection provides real-time access to critical structural information such as molecular weight, molecular formula, and diagnostic fragments—information that is necessary for quick online characterization and dereplication of NPs. The significant challenge in HPLC-MS used to be the clash between the high liquid flows from HPLC and the high vacuum needed for the mass spectrometer. Since the early 1980s, various interfaces have been developed to tackle this challenge and surmount these obstacles [38]. However, they were plagued by numerous other drawbacks and eventually fell out of favor. A more contemporary "transport-type" interface in LC-MS is the particle beam [39]. Its use in the examination of natural goods has, however, been restricted. A novel wave of direct-EI interfaces combined with nano-HPLC seems to have promise in generating these spectra rich in information [40]. In applications of HPLC-MS, various types of mass spectrometers can be employed. Low-resolution (LR) instruments, such as single quadrupole (Q) mass spectrometers, are commonly chosen due to their cost-effectiveness and widespread use. Conversely, high-resolution (HR) instruments with exact mass capabilities, such as time-of-flight (TOF) devices, are gaining popularity. They enhance certain detection capabilities by providing real-time molecular formula information and extremely accurate ion trace extraction (±0.01 Da). Triple-quadrupole (QQQ) MS-MS systems are often utilized for highly specific detection or to analyze structurally relevant fragments. These devices are especially useful for multiple reaction monitoring (MRM) techniques-based precision quantification in complicated matrices and bioanalytical tests [41]. The distinctive feature of ion-trap mass spectrometers is their capability to generate multiple-stage MS-MS (MS^n) data, which can be extremely valuable during the process of structural elucidation. In addition to these four categories of mass spectrometers, an increasing number of other models are available, such as hybrid systems that combine high- and low-resolution (HR) analyzers for specific applications [38, 42].

3.10 HPLC-NMR (nuclear magnetic resonance)

Real-time structural information is indeed the primary application of HPLC-NMR. However, HPLC-NMR can also be considered a universal detection technique because 1H-NMR can identify all protons associated with analytes [43]. However, its limited sensitivity and significant implementation costs are the main reasons it is not used for this purpose. Fluorinated derivatives have been selectively detected using HPLC-19F-NMR. An instance of this occurred during the investigation of 3-chloro-4-fluoroaniline's metabolic routes in rat bodily fluids [44].

4. Detection methods for online structural identification

To provide structurally relevant information for the online identification and dereplication of natural products (NPs), HPLC can be integrated with various spectroscopic methods used for NP detection. These methods include DAD, MS, NMR, CD, and IR. Techniques that use both spectroscopic and chromatographic dimensions are referred to as "LC-hyphenated techniques." Using these techniques to dereplicate NPs is a strategic way to conduct an efficient lead-finding procedure [45]. Unfortunately, the absence of comprehensive LC spectroscopic databases makes this task challenging to accomplish. Moreover, even if libraries of natural product spectra could be efficiently generated on a particular LC-hyphenated system, the reliance of much of the data, particularly HPLC-MS and HPLC-MS-MS, on the specific instrument used hampers their broad acceptance, unlike GC-MS [46]. As a result, the atom configuration and stereochemical orientations for every new compound have to be established from the beginning. As a result, no one LC-hyphenated method exists that is capable of identifying every secondary metabolite present in an online plant extract [47]. The goal of hyphenated procedures should be to provide spectroscopic data that is on par with that of isolated, pure products. This is untrue, though, and every method has its limitations. Further adjustments are necessary when combining different hyphenated techniques (hyphenation), such as ensuring compatibility of eluent compositions, sensitivity, and dynamic range of the methods being hyphenated. Nonetheless, it offers the advantage of presenting all structural information in a single analysis [48]. If these methods fail to replicate unknown natural compounds, supplementary information can be acquired through NMR, either directly linked with LC (HPLC-NMR) [49] or is utilized in conjunction with preconcentration techniques like HPLC-SPE-NMR [50, 51] or after micro fractionation with microflow HPLC-NMR methods such as Cap NMR [45], or even with a combination of both techniques, such as CAP-SPE-NMR [52]. Additional hyphenated methods, like HPLC-IR [53] or HPLC-CD [54], can provide important extra structural details.

4.1 HPLC-DAD (photodiode array detection)

Photodiode array detection (DAD) provides UV spectra directly online, offering particular advantages in detecting natural products with characteristic chromophores [55]. For example, this approach works well for identifying polyphenols since they have unique chromophores. DAD-UV spectral libraries can be created and used for dereplication using these kinds of molecules. Compound analysis must be done under constant HPLC conditions, nevertheless, as fluctuations in the composition of the mobile phase may have a modest impact on the UV bands. Moreover, the selection of modifiers could impact UV detection and potentially obscure low-wavelength bands. Peak purity can be assessed by obtaining numerous UV spectra spanning a certain LC peak [56]. HPLC-DAD analysis has been used exclusively to partially characterize a large number of NPs with distinctive chromophores, mostly phenolics [56], polyketides, alkaloids, and terpenoids [56]. Furthermore, this method has been effectively utilized alongside UV shift reagents commonly employed for the structural characterization of pure flavonoids [46, 57]. The accurate location of hydroxy groups on the polyphenols online can be achieved by comparing the original and shifted online DAD spectra. The ability to simultaneously monitor several wavelengths for the identification of various classes of chemicals is another noteworthy feature of DAD-UV detection. This is because all wavelengths are captured during analysis. This ability has shown to be very helpful in simultaneously detecting the phloroglucinols (270 nm) and naphthodianthrones (590 nm) in *Hypericum perforatum* [58]. On the other hand, it allows for the reasonably selective measurement of anthocyanidins (520 nm), flavonoids and cinnamic derivatives (325 nm), indolic derivatives (270 nm), and glucosinolates (227 nm) in different Brassicaceae species [59]. Regarding basic HPLC-UV, there are many different uses for HPLC-DAD in the examination of natural goods [47, 60], and a thorough examination of them is outside the purview of this article. In multi-hyphenated systems, HPLC-DAD systems are frequently paired with MS or ELSD detection techniques [61, 62].

4.2 HPLC-MS, MS-MS, and MSn

4.2.1 Mass spectrometry, tandem mass spectrometry, and multiple-stage mass spectrometry

An essential method for the online identification of natural materials is HPLC-MS. Apart from its primary function of detection, a mass spectrometer can also provide nominal mass molecular ions or precise mass measurements that can be utilized to establish empirical formulas [43]. Additionally, the use of tandem or hybrid MS equipment makes it easier to collect precise structural data by breaking the molecular species apart via collision-induced dissociation (CID) reactions [38]. A number of reviews explore features of online identification of natural products for screening and dereplication [46, 63, 64]. These reviews cover a range of topics, including quality control, fingerprinting for authentication or standardization [38], and the identification of biomarkers [65]. Finding the molecular weight is essential for online dereplication. To distinguish between protonated or deprotonated molecules and adducts or fragments, this process requires comparing MS data acquired under different detection settings [55]. Utilizing high-resolution tools, like a TOF-MS (time-of-flight mass spectrometry) [66, 67] or FT-MS (Fourier-transform mass spectrometry) [60, 68] system, allows the molecular formula of crude mixes to be directly determined. This strategic information aids in the dereplication process, enabling a more targeted search within natural product libraries [46].

4.3 HPLC-NMR (nuclear magnetic resonance)

Due to the lack of comprehensive commercial MS-MS databases, additional spectroscopic data is often necessary throughout the dereplication process to validate the identity of known natural products or partially identify novel metabolites. In this regard, HPLC-NMR serves as a potent tool capable of providing crucial supplementary data or, in certain instances, facilitating a real-time structural elucidation of natural products in their entirety [46, 50, 69, 70]. With a well-resolved corresponding LC peak (noise peak level), HPLC-NMR should offer comprehensive structural characterization of any plant metabolite directly in an extract. In actuality, though, this statement is not totally true because a number of factors make it difficult to determine the internet structure [71]. The challenges primarily arise from the necessity for solvent suppression and the inherently low sensitivity of NMR in detecting microgram or sub-microgram quantities of natural compounds separated by conventional HPLC.

5. Applications of high-performance liquid chromatography in the analysis of herbal products

5.1 Recent advances in separation and analysis of saponins in natural products

Researchers have determined that ionic liquids and high-performance countercurrent chromatography are the predominant methods for extracting and separating saponins. Additionally, the combined chromatography technique is widely acknowledged as the primary method for analyzing saponins. Liquid chromatography, when paired with various detectors, offers the capability to conduct qualitative or quantitative analyses as well as quality control assessments of saponin compounds in medicinal materials and their preparations. In recent times, HPLC has gained extensive use for the identification and quantitative analysis of saponins and their preparations due to its exceptional resolution, selectivity, and sensitivity. The integration of HPLC with diverse detectors has emerged as the primary approach for saponin analysis. Prominent combinations include UV/diode array detectors, evaporative light-scattering detectors (ELSD), charged aerosol detectors (CAD), chromatographic fingerprinting, and mass spectrometer detector (MS). LC is coupled with these detectors to perform qualitative or quantitative analyses of multiple saponins present in complex medicinal materials and their preparations [72].

5.2 Role of HPLC in enhancing product quality and safety through quality control practice

For centuries, herbal medicinal products have served as a cornerstone of healthcare in various cultures around the globe [73, 74]. These products, which are generated from plants, offer a plethora of bioactive chemicals that have the potential to be used in therapeutic settings [75]. Ayurveda, traditional Chinese medicine (TCM), and native healing methods are just a few of the many traditional systems that use herbal medications [76]. The perception of natural origin and historical use of herbal medicinal items is a major reason for their continued popularity. Many communities have longstanding customs of using herbal remedies to treat a wide range of illnesses and improve general health [77]. This traditional wisdom has been passed down through the years, offering priceless insights into the therapeutic qualities of many plant species [78]. Products made from herbal medicines have demonstrated therapeutic promise in treating a variety of illnesses, from respiratory and digestive issues to chronic pain and immune system support [79]. Alkaloids, flavonoids, terpenes, and polyphenols are among the bioactive ingredients in herbal medicine products that have the capacity to interact with biological systems and provide possible therapeutic advantages [80]. However, questions about efficacy, safety, and quality control are also raised by the growing demand for herbal medicine products [81]. Ensuring the consistent quality and standardization of herbal products is essential to ensuring their safety, efficacy, and reproducibility [82]. Enforcing strict quality control procedures is essential to protecting consumers' health and well-being [83]. Herbal medicine products are safe to use since possible dangers like adulteration, contamination, and the presence of hazardous compounds can be reduced by putting strict testing and quality assurance procedures in place [84]. Efficient quality control methods involve evaluating procedures and identifying areas that need enhancement [71]. By optimizing processes, getting rid of obstacles, and fixing inefficiencies, businesses can increase production and productivity while consuming fewer resources [71]. The first step involves creating a uniform method for classifying herbs and guaranteeing efficient quality control procedures. This procedure entails keeping active ingredients or indicators at steady, reliable concentrations in herbal medicines [85]. Its objective is to minimize variations among batches and ensure that each product meets predetermined standards for quality [86]. Determining the primary active compounds or markers within the herb responsible for its therapeutic properties involves several critical elements. These include identifying active compounds, conducting quantitative analyses, and establishing reference standards. These tasks can be accomplished through scientific investigation, traditional wisdom, or by consulting existing literature [87]. Create techniques for accurately measuring the concentrations of active compounds or markers in quantitative terms [88]. Such methods may utilize chromatographic techniques like HPLC or GC, spectroscopic methods such as UV-Vis or IR, or dedicated chemical assays [89]. High-performance liquid chromatography, or HPLC, is a reliable analytical technique used to detect and separate the chemical components found in herbs [90]. Through the identification and measurement of particular marker chemicals or active components, HPLC confirms their existence and concentration, ensuring the consistency and quality of herbal medicine products [91]. Herbal species identification and authentication are accomplished molecularly with the use of DNA testing, namely DNA barcoding [92]. DNA barcoding involves sequencing a specific portion of the herb's DNA, such as the barcode region, which demonstrates species-specific uniqueness. By examining the acquired DNA sequence, the genetic identity of the herb can be determined through comparison with a reference database. This procedure helps to identify any possible adulteration or substitution and guarantees that the right species is used [93]. Herb identification can be done with great reliability using DNA testing, which is especially useful for processed or powdered herbs. Analytical techniques like nuclear magnetic resonance (NMR), gas chromatography-mass spectrometry (GC-MS), and high-performance liquid chromatography (HPLC) are used in chemical fingerprinting to analyze the chemical makeup of herbs [94]. Herb manufacturers can verify the authenticity, consistency, and quality of their products by comparing the chemical compositions of a sample of the herb with reference profiles that have been created. Chemical fingerprinting helps detect certain marker molecules or active components, ensuring that herbal medicine products have the appropriate potency and therapeutic efficacy [95]. For quality monitoring of herbal medicine products, objective, scientifically based data are provided by chemical fingerprinting and DNA testing [96]. Manufacturers can ensure the safety and consistency of herbal medication products, meet regulatory requirements, establish robust standards, maintain consistency, and detect adulterations or contaminants by incorporating these techniques into their quality control processes [73].

5.3 Analysis of herbal products by high performance liquid chromatography

5.3.1 Specific examples of step-by-step protocols: Single and multiple components analysis using markers

Various authors have extensively reviewed the utilization of HPLC methods in analyzing herbal products, and in recent years, a substantial body of literature on this subject has emerged. The following sections will provide specific examples, along with detailed protocols, showcasing the analysis process [97].

5.3.1.1 Liquorice (Glycyrrhiza glabra)

Liquorice (*Glycyrrhiza glabra*) serves as a clear example for protocols in HPLC analysis of herbal products, using marker compounds. Liquorice, derived from the roots of *Glycyrrhiza glabra*, is a well-known herbal medicinal product, with glycyrrhizin as its primary active component, typically ranging from 2% to 15% w/w. Additionally, liquorice is used as a natural sweetener due to the presence of glycyrrhizin, which is 200 times sweeter than sucrose. Dried liquorice roots are commercially available both as an herbal product and as a food additive. A semi-preparative HPLC-PDA method, employing glycyrrhizin as the marker compound, has been recently published [98]. Below is a summary of the protocol:

1. Nine liquorice samples were procured from worldwide marketplaces.

2. Ground samples (15 g each) were extracted using Soxhlet using 400 mL of MeOH and n-hexane in turn. The extracts were dried by filtering and evaporating them in a vacuum. Prior to HPLC analysis, the residue from each sample's extracts was redissolved in 6.05 mL of MeOH and filtered (or TLC in the case of n-hexane extracts). Since there was no glycyrrhizin in the n-hexane extract (as shown by TLC), HPLC analysis was performed on the MeOH extracts of all nine samples.

3. The reference marker was 20 mg/mL of glycyrrhizin in MeOH. MeOH was then used to make additional dilutions of 15, 10, 5, and 1 mg/mL. The calibration curve was created by injecting each stock dilution into the HPLC in triplicate.

4. Each methanol (MeOH) extract (100 mL) underwent analysis using an optimized and validated semi-preparative HPLC-PDA method. This method utilized a reversed-phase ACE 10C18-HL column (150 × 10 mm, 10 μm) with a C18 guard column ACE3310110GD (10 × 10 mm, 10 μm). The detection/monitoring wavelength was set at 254 nm, and the column temperature was maintained at 25°C. A binary gradient solvent system (30–100% B in A over 30 minutes) was employed, with a flow rate of 3.00 mL/minutes. Solvent A consisted of 0.1% v/v trifluoroacetic acid (TFA) in water, while solvent B was 0.1% v/v TFA in methanol (MeOH). The analysis was conducted using a Dionex 3000 LC System equipped with Dionex 3000 semi-preparative pumps, a Dionex 3000 autosampler, a Dionex 3000 column chamber, and a Dionex PDA-3000 detector. Data analysis was performed using the Chromeleon® Chromatography Data System.

5. Peak area was used to measure the amount of glycyrrhizin contained in each sample, while retention time and UV spectra were utilized to identify glycyrrhizin. The samples' glycyrrhizin content ranged from 0.177% to 0.688% w/w of dry materials.

5.3.2 Specific examples of step-by-step protocols: Chemical fingerprinting

The HPLC-based fingerprinting of different herbal products, including both mono- and polyherbals, is demonstrated by a number of instances found in the literature. While HPLC-PDA seems to be the most widely used method, probably because it is the most affordable and straightforward, there are numerous other sophisticated hyphenated approaches as well, such as LC-PDA-MS, that are also used [99, 100]. For instance, the qualitative and quantitative studies of Compound Kushen Injection, a well-known Chinese herbal product, were conducted using both HPLC and LC-DAD-MS/MS techniques [101]. Using a simple fingerprinting methodology, the study selected eight peaks from the HPLC chromatograms of 27 different batches of the herbal product to serve as characteristic peaks. Peak identification was carried out using LC-MS/MS and HPLC analyses. By comparing retention durations and UV spectra with those of genuine chemicals and by using summary MS fragmentation rules, a total of 21 chromatographic peaks were found. There were variations in the overall levels of marker chemicals among the several batches of this product. An example of the successive processes needed to chemically print herbal products.

5.3.2.1 Curculiginis Rhizoma (Curculigo orchioides)

The dried rhizome of *C. orchioides*, known as Curculiginis Rhizoma, is a popular Chinese herbal remedy and a significant Ayurvedic product that is used to treat inflammatory conditions, pain, hepatic and immunological problems, and as an aphrodisiac [102]. It is thought that curculigoside, a phenolic glycoside, plays a vital role in this herbal product's therapeutic qualities. However, the author has recently published on the use of a chemical fingerprinting method based on HPLC for evaluating the quality of this herbal product. This is due to the widely acknowledged fact that any herbal product's overall therapeutic activity (and/or toxicity) is typically the consequence of the synergistic action of several different chemicals present in the product rather than the sole cause of the activity [102]. The protocol is summarized in the section below.

1. Ten distinct Curculiginis Rhizoma samples were gathered from China.

2. After refluxing each sample (2 g) for 60 minutes with 45 mL of MeOH, the resultant mass was centrifuged for 15 minutes at 3000 rpm.

3. After the supernatant was poured into a 50 mL volumetric flask and filled with MeOH to reach the 50 mL threshold, it was filtered over a 0.45 μm cellulose acetate membrane in preparation for HPLC analysis.

4. Every sample (10 μL) was analyzed using an HPLC-PDA method that has been optimized and confirmed. We used an Altima C18 ODS column (4.6 × 250 mm, 5 mm) with a 30°C temperature setting. As solvents A and B, respectively, acetonitrile with 0.02% TFA and water with 0.5% TFA were employed in a step-wise gradient (7–30% A in B over 35 minutes) at a flow rate of 1.0 mL/minutes. The G1311A Quatpump solvent delivery system, UV-Vis photodiode array detector G1315B, and degasser G1322A were all included in the Agilent 1100 liquid chromatographic system that was utilized.

5. To find similarities between different chromatograms, the State Food and Drug Administration of China (SFDA) recommended utilizing the Similarity Evaluation System for Chromatographic Fingerprint of Traditional Chinese Medicine (Version 2004A) for similarity analysis. The similarity study was conducted using the correlation coefficient as a metric.

6. Eleven of the twenty distinct peaks were identified as "common peaks" since they were present in every sample.

6. Conclusion

Recently, herbal medicinal products represent an important source of drugs for different diseases; they contain a mixture of active therapeutic constituents, but they cannot be used as crude substances. Different methods are used to separate and identify such constituents in order to evaluate their specific activity and concentration. In addition, as this herbal constituent could be used as a drug, it will undergo process quality control steps. High-performance liquid chromatography (HPLC), and

DOI: http://dx.doi.org/10.5772/intechopen.1007159

Spectrophotometric methods represent a reliable analytical technique used to detect and separate the chemical components found in herbs, as well as the identification, purification, quality control, and standardization of such herbal constituents as they are combined with different detectors and advanced procedures. Multiple detection techniques are employed simultaneously, such as LC-UV-Vis-MS, LC-MS-MS, and LC-NMR-MS, to obtain comprehensive information. Thin layer chromatography (TLC) and high-performance thin layer chromatography (HPTLC) are valuable tools for qualitatively determining small amounts of impurities. In addition, high-performance liquid chromatography (HPLC) is used in chemical fingerprinting to analyze the chemical makeup of herbs. Recent advanced applications are demonstrating the usefulness of HPLC methods in analyzing herbal products.

Author details

Ali Gamal Al-Kaf[1,2]*, Rana Abdullah Al-Robaidi[1] and Hend Ahmed Al-Haj[3]

1 Medicinal Chemistry Department, Faculty of Pharmacy, Sana'a University, Yemen

2 Department of Pharmacy, Faculty of Medical Sciences, National University, Yemen

3 Pharmacognosy Department, Faculty of Pharmacy, Sana'a University, Yemen

*Address all correspondence to: alialkaf21@gmail.com; alialkaf@national-univ.net

References

[1] Sekhon BS. Inorganics/bioinorganics: Biological, medicinal and pharmaceutical uses. Journal of Pharmaceutical Education and Research. 2011;**2**(1):1

[2] Kushwaha SK, Kushwaha N, Maurya N, Rai A. Role of markers in the standardization of herbal drugs: A review. Archives of Applied Science Research. 2010;**2**(1):225-229

[3] Kamboj A. Analytical evaluation of herbal drugs. In: Drug Discovery Research in Pharmacognosy. Vol. 3. Landran, Mohali, India: Chandigarh College of Pharmacy; 2012. pp. 23-55

[4] Farooqui NA, Dey A, Singh G, Easwari T, Pandey MK. Analytical techniques in quality evaluation of herbal drugs. Asian Journal of Pharmaceutical Research. 2014;**4**(3):112-117

[5] Larsson J, Gottfries J, Muresan S, Backlund A. ChemGPS-NP: Tuned for navigation in biologically relevant chemical space. Journal of Natural Products. 2007;**70**(5):789-794

[6] Natishan TK. Recent developments of achiral HPLC methods in pharmaceuticals using various detection modes. Journal of Liquid Chromatography and Related Technologies. 2004;**27**(7-9):1237-1316

[7] Nguyen DTT, Guillarme D, Rudaz S, Veuthey JL. Fast analysis in liquid chromatography using small particle size and high pressure. Journal of Separation Science. 2006;**29**(12):1836-1848

[8] François I, Tienpont B, David F, Vanhoenacker G, Sandra P. Coupling Columns and Multidimensional Configurations to Increase Peak Capacity in Liquid Chromatography. LCGC Europe. LCGC Europe-03-01-2007. 2007;**20**(3):154-162

[9] Wilson I, Brinkman UT. Hyphenation and hypernation: The practice and prospects of multiple hyphenation. Journal of Chromatography A. 2003;**1000**(1-2):325-356

[10] Wolfender J-L. HPLC in natural product analysis: The detection issue. Planta Medica. 2009;**75**(07):719-734

[11] Scott RP. Principles and Practice of Chromatography. Chrom-ed Book Series. Vol. 12003

[12] Gao X, Dan M, Zhao A, Xie G, Jia W. Simultaneous determination of saponins in flower buds of *Panax notoginseng* using high performance liquid chromatography. Biomedical Chromatography. 2008;**22**(3):244-249

[13] Özkan S. LC with electrochemical detection. Recent application to pharmaceuticals and biological fluids. Chromatographia. 2007;**66**(Suppl 1):3-13

[14] Leitner A, Emmert J, Boerner K, Lindner W. Influence of solvent additive composition on chromatographic separation and sodium adduct formation of peptides in HPLC–ESI MS. Chromatographia. 2007;**65**:649-653

[15] Braga SMLFM, de Medeiros FD, de Jesus OE, Macedo RO. Development and validation of a method for the quantitative determination of aflatoxin contaminants in *Maytenus ilicifolia* by HPLC with fluorescence detection. Phytochemical Analysis: An International Journal of Plant Chemical and Biochemical Techniques. 2005;**16**(4):267-271

[16] Klvana M, Chen J, Lépine F, Legros R, Jolicoeur M. Analysis of secondary metabolites from eschscholtzia californica by high-performance liquid chromatography. Phytochemical Analysis: An International Journal of Plant Chemical and Biochemical Techniques. 2006;**17**(4):236-242

[17] Li F, Zhang C, Guo X, Feng W. Chemiluminescence detection in HPLC and CE for pharmaceutical and biomedical analysis. Biomedical Chromatography. 2003;**17**(2-3):96-105

[18] Ohba Y, Kuroda N, Nakashima K. Liquid chromatography of fatty acids with chemiluminescence detection. Analytica Chimica Acta. 2002;**465**(1-2):101-109

[19] Zhang Q, Cui H. Simultaneous determination of quercetin, kaempferol, and isorhamnetin in phytopharmaceuticals of *Hippophae rhamnoides* L. by high-performance liquid chromatography with chemiluminescence detection. Journal of Separation Science. 2005;**28**(11):1171-1178

[20] LaCourse WR, Modi SJ. Microelectrode applications of pulsed electrochemical detection. Electroanalysis: An International Journal Devoted to Fundamental and Practical Aspects of Electroanalysis. 2005;**17**(13):1141-1152

[21] Eggelkraut-Gottanka S, Abed SA, Müller W, Schmidt P. Quantitative analysis of the active components and the by-products of eight dry extracts of *Hypericum perforatum* L. (St John's Wort). Phytochemical Analysis: An International Journal of Plant Chemical and Biochemical Techniques. 2002;**13**(3):170-176

[22] Ding C, Chen E, Zhou W, Lindsay RC. A method for extraction and quantification of Ginkgo terpene trilactones. Analytical Chemistry. 2004;**76**(15):4332-4336

[23] Megoulas NC, Koupparis MA. Twenty years of evaporative light scattering detection. Critical Reviews in Analytical Chemistry. 2005;**35**(4):301-316

[24] Guillarme D, Rudaz S, Schelling C, Dreux M, Veuthey J-L. Micro liquid chromatography coupled with evaporative light scattering detector at ambient and high temperature: Optimization of the nebulization cell geometry. Journal of Chromatography A. 2008;**1192**(1):103-112

[25] Dubber M-J, Kanfer I. Determination of terpene trilactones in *Ginkgo biloba* solid oral dosage forms using HPLC with evaporative light scattering detection. Journal of Pharmaceutical and Biomedical Analysis. 2006;**41**(1):135-140

[26] Li P, Zeng L, Li S, Lin G. The extraction of imperialine and imperialine-3β-glucoside from *Fritillaria pallidiflora* Schrenk and quantitative determination by HPLC-evaporative light scattering detection. Phytochemical Analysis: An International Journal of Plant Chemical and Biochemical Techniques. 2002;**13**(3):158-161

[27] Kim SN, Ha YW, Shin H, Son SH, Wu SJ, Kim YS. Simultaneous quantification of 14 ginsenosides in Panax ginseng CA Meyer (Korean red ginseng) by HPLC-ELSD and its application to quality control. Journal of Pharmaceutical and Biomedical Analysis. 2007;**45**(1):164-170

[28] Schaneberg BT, Molyneux RJ, Khan IA. Evaporative light scattering detection of pyrrolizidine alkaloids. Phytochemical Analysis: An International Journal of Plant Chemical and Biochemical Techniques. 2004;**15**(1):36-39

[29] Cremin PA, Zeng L. High-throughput analysis of natural product compound libraries by parallel LC-MS evaporative light scattering detection. Analytical Chemistry. 2002;**74**(21):5492-5500

[30] Lísa M, Lynen F, Holčapek M, Sandra P. Quantitation of triacylglycerols from plant oils using charged aerosol detection with gradient compensation. Journal of Chromatography A. 2007;**1176**(1-2):135-142

[31] Dixon RW, Peterson DS. Development and testing of a detection method for liquid chromatography based on aerosol charging. Analytical Chemistry. 2002;**74**(13):2930-2937

[32] Vervoort N, Daemen D, Török G. Performance evaluation of evaporative light scattering detection and charged aerosol detection in reversed phase liquid chromatography. Journal of Chromatography. A. 2008;**1189**(1-2):92-100

[33] Cole RO, Freeto SM, Gamache PH, Woodcock MJ, McCarthy RS, Laws K, et al. HPLC Analysis of Nonvolatile Analytes Using Charged Aerosol Detection. LCGC North America. LCGC North America-02-01-2005. 2005;**23**(2):150-161

[34] Guillarme D, Heinisch S, Gauvrit J, Lanteri P, Rocca J. Optimization of the coupling of high-temperature liquid chromatography and flame ionization detection: Application to the separations of alcohols. Journal of Chromatography A. 2005;**1078**(1-2):22-27

[35] Smith RM. Superheated water chromatography—A green technology for the future. Journal of Chromatography A. 2008;**1184**(1-2):441-455

[36] Guillarme D, Heinisch S. Detection modes with high temperature liquid chromatography—A review. Separation and Purification Reviews. 2005;**34**(2):181-216

[37] Cheng K-W, Chen F, Wang M. Liquid chromatography-mass spectrometry in natural product research. In: Colegate SM, Russell J, editors. Molyneux. Bioactive Natural Products. Boca Raton: CRC Press; 2007. pp. 259-280

[38] Niessen WM. Liquid Chromatography-Mass Spectrometry. CRC Press; 2006

[39] Brewer TM, Castro J, Marcus RK. Particle beam sample introduction into glow discharge plasmas for speciation analysis. Spectrochimica Acta Part B: Atomic Spectroscopy. 2006;**61**(2):134-149

[40] Cappiello A, Famiglini G, Pierini E, Palma P, Trufelli H. Advanced liquid chromatography-mass spectrometry interface based on electron ionization. Analytical Chemistry. 2007;**79**(14):5364-5372

[41] Pisitkun T, Hoffert JD, Yu M-J, Knepper MA. Tandem mass spectrometry in physiology. Physiology. 2007;**22**(6):390-400

[42] Korfmacher WA. Foundation review: Principles and applications of LC-MS in new drug discovery. Drug Discovery Today. 2005;**10**(20):1357-1367

[43] Wolfender JL, Ndjoko K, Hostettmann K. The potential of LC-NMR in phytochemical analysis. Phytochemical Analysis: An International Journal of Plant Chemical and Biochemical Techniques. 2001;**12**(1):2-22

[44] Duckett C, Lindon J, Walker H, Abou-Shakra F, Wilson I, Nicholson J. Metabolism of 3-chloro-4-fluoroaniline in rat using [14C]-radiolabelling,

19F-NMR spectroscopy, HPLC-MS/MS, HPLC-ICPMS and HPLC-NMR. Xenobiotica. 2006;**36**(1):59-77

[45] Wolfender J-L, Queiroz EF, Hostettmann K. The importance of hyphenated techniques in the discovery of new lead compounds from nature. Expert Opinion on Drug Discovery. 2006;**1**(3):237-260

[46] Waridel P, Wolfender J-L, Ndjoko K, Hobby KR, Major HJ, Hostettmann K. Evaluation of quadrupole time-of-flight tandem mass spectrometry and ion-trap multiple-stage mass spectrometry for the differentiation of C-glycosidic flavonoid isomers. Journal of Chromatography A. 2001;**926**(1):29-41

[47] Wolfender J-L, Queiroz EF, Hostettmann K. Development and application of LC-NMR techniques to the identifi cation of bioactive natural products. In: Bioactive Natural Products. CRC Press; 2007. pp. 157-204

[48] Wilson ID, Udo A. Hype and hypernation: Multiple hyphenation of column liquid chromatography and spectroscopy. TrAC Trends in Analytical Chemistry. 2007;**26**(9):847-854

[49] Jaroszewski JW. Hyphenated NMR methods in natural products research, part 1: Direct hyphenation. Planta Medica. 2005;**71**(08):691-700

[50] Jaroszewski JW. Hyphenated NMR methods in natural products research, part 2: HPLC-SPE-NMR and other new trends in NMR hyphenation. Planta Medica. 2005;**71**(09):795-802

[51] Clarkson C, Madikane EV, Hansen SH, Smith PJ, Jaroszewski JW. HPLC-SPE-NMR characterization of sesquiterpenes in an antimycobacterial fraction from Warburgia salutaris. Planta Medica. 2007;**73**(06):578-584

[52] Lambert M, Wolfender J-L, Stærk D, Christensen SB, Hostettmann K, Jaroszewski JW. Identification of natural products using HPLC-SPE combined with CapNMR. Analytical Chemistry. 2007;**79**(2):727-735

[53] Louden D, Handley A, Taylor S, Lenz E, Miller S, Wilson ID, et al. Spectroscopic characterisation and identification of ecdysteroids using high-performance liquid chromatography combined with on-line UV–diode array, FT-infrared and 1H-nuclear magnetic resonance spectroscopy and time of flight mass spectrometry. Journal of Chromatography A. 2001;**910**(2):237-246

[54] Ren D-M, Guo H-F, Wang S-Q, Lou H-X. Separation and structure determination of two diastereomeric pairs of enantiomers from *Dracocephalum rupestre* by high-performance liquid chromatography with circular dichroism detection. Journal of Chromatography. A. 2007;**1161**(1-2):334-337

[55] Larsen TO, Hansen MA. Dereplication and discovery of natural products by UV spectroscopy. In: Bioactive Natural Products. CRC Press; 2007. pp. 235-258

[56] Robbins RJ. Phenolic acids in foods: An overview of analytical methodology. Journal of Agricultural and Food Chemistry. 2003;**51**(10):2866-2887

[57] Waridel P, Wolfender J-L, Lachavanne J-B, Hostettmann K. Identification of the polar constituents of Potamogeton species by HPLC-UV with post-column derivatization, HPLC-MSn and HPLC-NMR, and isolation of a new ent-labdane diglycoside. Phytochemistry. 2004;**65**(16):2401-2410

[58] Tolonen A, Hohtola A, Jalonen J. Fast high-performance liquid chromatographic analysis of naphthodianthrones and phloroglucinols

from *Hypericum perforatum* extracts. Phytochemical Analysis: An International Journal of Plant Chemical and Biochemical Techniques. 2003;**14**(5):306-309

[59] Bennett RN, Mellon FA, Foidl N, Pratt JH, Dupont MS, Perkins L, et al. Profiling glucosinolates and phenolics in vegetative and reproductive tissues of the multi-purpose trees *Moringa oleifera* L. (horseradish tree) and *Moringa stenopetala* L. Journal of Agricultural and Food Chemistry. 2003;**51**(12):3546-3553

[60] Suzuki H, Sasaki R, Ogata Y, Nakamura Y, Sakurai N, Kitajima M, et al. Metabolic profiling of flavonoids in *Lotus japonicus* using liquid chromatography Fourier transform ion cyclotron resonance mass spectrometry. Phytochemistry. 2008;**69**(1):99-111

[61] Gobbo-Neto L, Lopes NP. Online identification of chlorogenic acids, sesquiterpene lactones, and flavonoids in the Brazilian arnica *Lychnophora ericoides* Mart. (Asteraceae) leaves by HPLC-DAD-MS and HPLC-DAD-MS/MS and a validated HPLC-DAD method for their simultaneous analysis. Journal of Agricultural and Food Chemistry. 2008;**56**(4):1193-1204

[62] Li J, Qi H, Qi L-W, Yi L, Li P. Simultaneous determination of main phytoecdysones and triterpenoids in radix achyranthis bidentatae by high-performance liquid chromatography with diode array-evaporative light scattering detectors and mass spectrometry. Analytica Chimica Acta. 2007;**596**(2):264-272

[63] Sashidhara KV, Rosaiah JN. Various dereplication strategies using LC-MS for rapid natural product lead identification and drug discovery. Natural Product Communications. 2007;**2**(2):1934578X0700200218

[64] Yang M, Sun J, Lu Z, Chen G, Guan S, Liu X, et al. Phytochemical analysis of traditional Chinese medicine using liquid chromatography coupled with mass spectrometry. Journal of Chromatography A. 2009;**1216**(11):2045-2062

[65] Kussmann M, Affolter M, Nagy K, Holst B, Fay LB. Mass spectrometry in nutrition: Understanding dietary health effects at the molecular level. Mass Spectrometry Reviews. 2007;**26**(6):727-750

[66] Grata E, Boccard J, Guillarme D, Glauser G, Carrupt P-A, Farmer EE, et al. UPLC–TOF-MS for plant metabolomics: A sequential approach for wound marker analysis in *Arabidopsis thaliana*. Journal of Chromatography B. 2008;**871**(2):261-270

[67] Xie G, Plumb R, Su M, Xu Z, Zhao A, Qiu M, et al. Ultra-performance LC/TOF MS analysis of medicinal Panax herbs for metabolomic research. Journal of Separation Science. 2008;**31**(6-7):1015-1026

[68] Bonkanka CX, Smelcerovic A, Zuehlke S, Rabanal RM, Spiteller M, del Carmen S-MC. HPLC-MS analysis and anti-oedematogenic activity of *Hypericum grandifolium* Choisy (Hypericaceae). Planta Medica. 2008;**74**(07):719-725

[69] Wolfender JL, Queiroz EF, Hostettmann K. Phytochemistry in the microgram domain—A LC–NMR perspective. Magnetic Resonance in Chemistry. 2005;**43**(9):697-709

[70] Yang Z. Online hyphenated liquid chromatography–nuclear magnetic resonance spectroscopy–mass spectrometry for drug metabolite and nature product analysis. Journal of Pharmaceutical and Biomedical Analysis. 2006;**40**(3):516-527

[71] Gong L, Xie J-b, Luo Y, Qiu Z-d, Liu J-r, Mei N-j, et al. Research progress of quality control for the seed of *Ziziphus jujuba* var. spinosa (Bunge) Hu ex HF Chow (Suan-Zao-Ren) and its proprietary Chinese medicines. Journal of Ethnopharmacology. 2023;**307**:116204

[72] Wang Y, Ma Y, Tao L, Zhang X, Hao F, Zhao S, et al. Recent advances in separation and analysis of saponins in natural products. Separations. 2022;**9**(7):163

[73] Wang H, Chen Y, Wang L, Liu Q, Yang S, Wang C. Advancing herbal medicine: Enhancing product quality and safety through robust quality control practices. Frontiers in Pharmacology. 2023;**14**:1265178

[74] Motti R. Wild plants used as herbs and spices in Italy: An ethnobotanical review. Plants. 2021;**10**(3):563

[75] Chrysant SG, Chrysant GS. Herbs used for the treatment of hypertension and their mechanism of action. Current Hypertension Reports. 2017;**19**:1-10

[76] Kabak B, Dobson AD. Mycotoxins in spices and herbs—An update. Critical Reviews in Food Science and Nutrition. 2017;**57**(1):18-34

[77] Alipour R, Marzabadi LR, Arjmand B, Ayati MH, Namazi N. The effects of medicinal herbs on gut microbiota and metabolic factors in obesity models: A systematic review. Diabetes and Metabolic Syndrome: Clinical Research and Reviews. 2022;**16**(9):102586

[78] Cho EC, Kim K. A comprehensive review of biochemical factors in herbs and their constituent compounds in experimental studies on alopecia. Journal of Ethnopharmacology. 2020;**258**:112907

[79] Shaikh AS, Thomas AB, Chitlange SS. Herb–drug interaction studies of herbs used in treatment of cardiovascular disorders—A narrative review of preclinical and clinical studies. Phytotherapy Research. 2020;**34**(5):1008-1026

[80] Brown AC. Liver toxicity related to herbs and dietary supplements: Online table of case reports. Part 2 of 5 series. Food and Chemical Toxicology. 2017;**107**:472-501

[81] Tao Y, Bao J, Zhu F, Pan M, Liu Q, Wang P. Ethnopharmacology of *Rubus idaeus* Linnaeus: A critical review on ethnobotany, processing methods, phytochemicals, pharmacology and quality control. Journal of Ethnopharmacology. 2023;**302**:115870

[82] Rahath Kubra I, Kumar D, Jagan Mohan Rao L. Emerging trends in microwave processing of spices and herbs. Critical Reviews in Food Science and Nutrition. 2016;**56**(13):2160-2173

[83] Deutch MR, Grimm D, Wehland M, Infanger M, Krüger M. Bioactive candy: Effects of licorice on the cardiovascular system. Food. 2019;**8**(10):495

[84] Giacometti J, Kovačević DB, Putnik P, Gabrić D, Bilušić T, Krešić G, et al. Extraction of bioactive compounds and essential oils from mediterranean herbs by conventional and green innovative techniques: A review. Food Research International. 2018;**113**:245-262

[85] Aleksieva KI, Yordanov ND. Various approaches in EPR identification of gamma-irradiated plant foodstuffs: A review. Food Research International. 2018;**105**:1019-1028

[86] Ketai W, Huitao L, Yunkun Z, Xingguo C, Zhide H, Yucheng S, et al. Separation and determination

of alantolactone and isoalantolactone in traditional Chinese herbs by capillary electrophoresis. Talanta. 2000;**52**(6):1001-1005

[87] Zhao N, Liu D, Wang Y, Zhang X, Zhang L. Screening and identification of anti-acetylcholinesterase ingredients from Tianzhi granule based on ultrafiltration combined with ultra-performance liquid chromatography-mass spectrometry and in silico analysis. Journal of Ethnopharmacology. 2022;**298**:115641

[88] An Y-l, Wei W-l, Guo D-a. Application of analytical technologies in the discrimination and authentication of herbs from Fritillaria: A review. Critical Reviews in Analytical Chemistry. 2024;**54**(6):1775-1796

[89] Pan M, Pei W, Yao Y, Dong L, Chen J. Rapid and integrated quality assessment of organic-inorganic composite herbs by FTIR spectroscopy—Global chemical fingerprints identification and multiple marker components quantification of indigo Naturalis (Qing Dai). Molecules. 2018;**23**(11):2743

[90] Sontag G, Pinto MI, Noronha JP, Burrows HD. Analysis of food by high performance liquid chromatography coupled with coulometric detection and related techniques: A review. Journal of Agricultural and Food Chemistry. 2019;**67**(15):4113-4144

[91] Malherbe CJ, De Beer D, Joubert E. Development of on-line high performance liquid chromatography (HPLC)-biochemical detection methods as tools in the identification of bioactives. International Journal of Molecular Sciences. 2012;**13**(3):3101-3133

[92] Li X, Yang Y, Henry RJ, Rossetto M, Wang Y, Chen S. Plant DNA barcoding: From gene to genome. Biological Reviews. 2015;**90**(1):157-166

[93] Senapati A, Basak S, Rangan L. A review on application of DNA barcoding technology for rapid molecular diagnostics of adulterants in herbal medicine. Drug Safety. 2022;**45**(3):193-213

[94] Li M, Xu J, Zheng Q, Guo C, Chen Y. Chemical-based surface plasmon resonance imaging of fingerprints. Analytical Chemistry. 2022;**94**(20):7238-7245

[95] Gainza P, Sverrisson F, Monti F, Rodola E, Boscaini D, Bronstein MM, et al. Deciphering interaction fingerprints from protein molecular surfaces using geometric deep learning. Nature Methods. 2020;**17**(2):184-192

[96] Yin F, Song Z, He Z, Qin B, John GF, Zhang L, et al. Chemical fingerprinting and characterization of spilled oils and burnt soot particles—A case study on the Sanchi oil tanker collision in the East China Sea. Science of the Total Environment. 2022;**824**:153896

[97] Hua-Bin Z, Ai-Qin D, Xin-Ling Z, Pei-Hai W, Wei-Jie L, Guo-Sheng Y, et al. Quality control methodology and their application in analysis on HPLC fingerprint spectra of herbal medicines. Chromatography Research International. 2011

[98] Basar N, Talukdar AD, Nahar L, Stafford A, Kushiev H, Kan A, et al. A simple semi-preparative reversed-phase HPLC/PDA method for separation and quantification of glycyrrhizin in nine samples of *Glycyrrhiza glabra* root collected from different geographical origins. Phytochemical Analysis. 2014;**25**(5):399-404

[99] Zhu H, Wang C, Qi Y, Song F, Liu Z, Liu S. Fingerprint analysis of Radix Aconiti using ultra-performance liquid chromatography–electrospray ionization/tandem mass spectrometry (UPLC–ESI/MSn) combined with stoichiometry. Talanta. 2013;**103**:56-65

[100] Zhang J-l, Cui M, He Y, Yu H-l, Guo D-a. Chemical fingerprint and metabolic fingerprint analysis of Danshen injection by HPLC–UV and HPLC–MS methods. Journal of Pharmaceutical and Biomedical Analysis. 2005;**36**(5):1029-1035

[101] Tang D-Q, Zheng X-X, Chen X, Yang D-Z, Du Q. Quantitative and qualitative analysis of common peaks in chemical fingerprint of Yuanhu Zhitong tablet by HPLC-DAD–MS/MS. Journal of Pharmaceutical Analysis. 2014;**4**(2):96-106

[102] Bian Q, Yang H, Chan C-o, Jin D, Mok DK-W, Chen S. Fingerprint analysis and simultaneous determination of phenolic compounds in extracts of Curculiginis Rhizoma by HPLC-diode array detector. Chemical and Pharmaceutical Bulletin. 2013;**61**(8):802-808

Chapter 9

A New HPLC-DAD Method for the Simultaneous Measurements of Glycyrrhizic Acid (GA) and Glabridin (GB) in Licorice (*Glycyrrhiza glabra* L.) Extract

Burhan Ceylan

Abstract

Glycyrrhiza glabra L. (Licorice) is a plant with important applications in the fields of medicine, food and cosmetics due to the extracts obtained from its root and the chemicals contained in these extracts. Glycyrrhizic acid (GA), a saponin glycosid, is the main component that gives licorice its sweet taste. The ammonium salt of this compound is listed in the European Pharmacopeia and is the sweetest natural flavoring listed on the Food and Drug Administration (FDA). Its main known properties are its effective properties in the treatment of ulcers, bronchitis, sore throats, liver diseases and its anti-inflammatory, antibacterial and antiviral effects (especially hepatitis C and HIV treatment). Glabridin (GB), an isoflavan, is known for its cytotoxic, antimicrobial, estrogenic and anti-proliferative effects in breast cancer. It is also valuable for its antioxidant effect on melanogenesis, low-density lipoprotein (LDL) oxidation and mitochondria functions. Due to all these effects, licorice is widely used by healthcare professionals, doctors and the public. It is extremely important to determine the contents of the active ingredients in licorice root extracts. This work devised an easy-to-use, fast and economical HPLC-DAD technique to analyze this bivalent chemical in products containing licorice root extract. The developed method will shed light on future bioavailability studies and the determination of these chemicals in extracts.

Keywords: *Glycyrrhiza glabra*, glycyrrhizic acid, glabridin, HPLC-DAD, method validation

1. Introduction

Glycyrrhiza glabra L. (Licorice) is a perennial woody shrub and is distributed in many regions of the Mediterranean and Asia. These include Turkey, Italy, Spain, Russia, Syria, Iran, China and Afghanistan [1]. The plant has blue-purple flowers. Licorice is a perennial shrub from the Fabaceae family that can grow up to

Figure 1.
General view and hammered root image of Glycyrrhiza glabra *L.*

approximately 120–150 cm tall. The root of the licorice plant is 2.5 ± 0.5 cm in diameter and 15–20 cm long. **Figure 1** shows the general appearance of the *Glycyrrhiza glabra* L. plant. This distribution is observed in our country in the provinces of Gaziantep, Iğdır, Kars, Siirt, Bitlis, Diyarbakir, Kahramanmaraş, Muş, Samsun and Şanlıurfa.

Licorice extract is 50 times sweeter than sugar and contains a mixture of triterpenoid saponins (4–20%) and potassium and calcium salts of glycyrrhizic acid (GA), including larger amounts of GA. Licorice root has a long history of medicinal use in Europe and Asia dating back thousands of years. It has a wider use in Asian countries and is used as a sweetener and also for healing wounds [2]. The syrup of this root is used effectively in the treatment of peptic ulcer disease, constipation, cough, bronchitis, cold, diabetes, lung and kidney stones [3]. In addition, it has a wide range of pharmacological effects such as antiulcer, antidiabetic, antiviral, antioxidant, antiallergic, anticancer, antidepressant, expectorant, anti-inflammatory in colitis and memory enhancer [4]. In addition to all these, it has been explained in the literature that licorice has uses in the treatment of osteoarthritis [5], systemic lupus erythematosus (SLE) [6], malaria [7], tuberculosis [8], food poisoning [9] and chronic fatigue syndrome (CFS) [10]. Shakuyaku-kanzo-to, a herbal product containing licorice, increases fertility in women with polycystic ovary syndrome [11, 12]. However, licorice has proven uses in the treatment of prostate cancer [13] and in the treatment of eczema [14, 15].

Some examples of products or preparations using this extract are cough syrups, food supplements, chewing gum, beverages and candies [16]. Tobacco product manufacturers have reported using licorice root to flavor tobacco and sweeten smoke at different stages of production. Licorice adds a ripe, sweet and woody flavor to the smoke. Additionally, in the cosmetic industry, licorice root can be used as a natural skin brightener as an alternative to chemical skin brighteners such as hydroquinone. In addition, it has been shown to be very beneficial to skin health when used externally, as it causes intracellular fat and hyaluronan synthesis [17].

The ammonium salt of GA (**Figure 2**) is listed in the European Pharmacopeia and is the sweetest natural flavor listed on the Food and Drug Administration (FDA). Most pharmacological studies on licorice focus on GA and its aglycone glycyrrhetic acid. These two compounds have many comprehensive biological activities such as antiulcer,

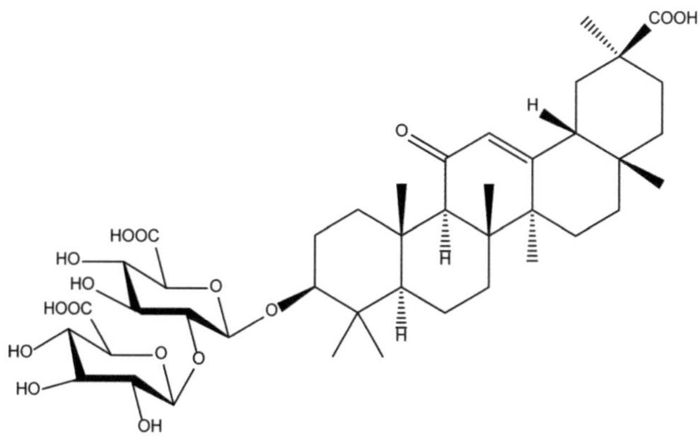

Figure 2.
Chemical structure of GA.

anti-inflammatory, antiallergic, antioxidant, anticarcinogenic, antithrombotic, antidiabetic and antiviral. Additionally, GA is used as a therapeutic agent against many viral diseases such as AIDS, and hepatitis B and C [18]. GA is a water-soluble pentacyclic triterpenoid glycoside that gives licorice its taste and is used as a sweetening additive and emulsifier in the food industry because it is 50 times sweeter than sugar. In many countries, GA is an important therapeutic agent used to treat allergic dermatitis and chronic viral hepatitis [2]. The acid form of GA is insoluble in water, but its ammonium salt is soluble in water (pH > 4.5). The monoammonium salt of GA is used as an anti-inflammatory, antiallergic agent for the treatment of bronchial asthma, eczema and other diseases [16]. In the cosmetic field, GA and stearyl glycyrrhetinate are used to reduce dryness and deformation in dry and deformed skin. It is also preferred after sun exposure due to its anti-inflammatory and soothing properties.

GB (**Figure 3**) has been reported to exhibit many pharmacological activities such as cytotoxic activity, antimicrobial activity and antiproliferative activity against breast cancer cells. Additionally, it is effective in protecting mitochondrial functions from melanogenesis, inflammation, low-density lipoprotein oxidation and oxidative stress [2]. It has been proven by research that GB significantly reduces inflammation in the intestine in colitis GB has an inhibitory effect on melanin production and inflammation [19]. It inhibits the formation of nitric oxide, a free radical. The use of GB at a concentration of 0.1% is sufficient to achieve the desired effect. It is also known to have an anti-inflammatory effect. It has a depigmentation effect that is 16 times stronger than hydroquinone, and no side effects have been detected in the long

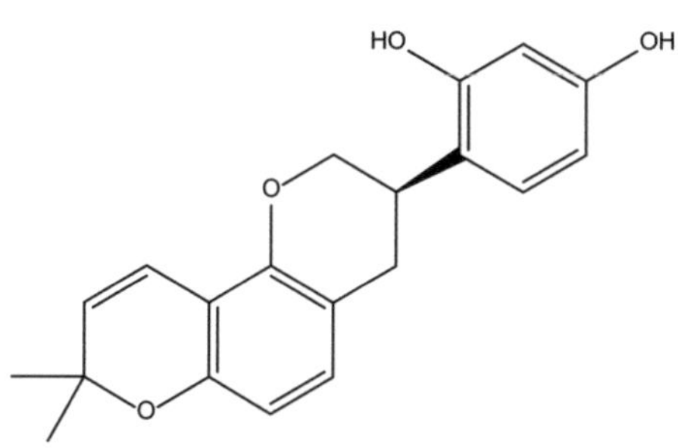

Figure 3.
Chemical structure of GB.

term [20]. GB is used as a whitening agent against age spots. GB helps heal red and scalded skin but is expensive for this treatment.

There are 2783 food supplements, cosmetic products or pharmaceutical formulations worldwide containing licorice extract or a pure compound containing this extract. While 116 of these contain high amounts of licorice, other products contain this extract as a sweetener or thickener in amounts up to 1% [21, 22]. FDA classifies licorice and its ingredients as generally recognized as safe (GRAS). This extract has not been described as toxic by any authority. When we look at the literature, there are very few methods for the determination of the components (GA and GB) responsible for the effect in products containing licorice extract (food supplements, cosmetics and drug formulations) using liquid chromatography. Current methods have been disadvantages due to the use of expensive detectors, long sample preparation procedures, and the small number of samples examined or the same type. Mass spectrometry is a unique analytical technique, but it is expensive when integrated with liquid chromatography. Additionally, in existing methods, the retention times of these two compounds (GA and GB) are quite long. Since the existing formulation in herbal extracts and food supplements is almost the same, there is no method that can determine such different types of formulations with a simple extraction method at low cost (UV detection) as in our study [23–25]. The aim of this study is to use body spray, cream, lotion, ointment, lozenge, nasal spray, nasal ointment, ear drops, soap and shampoo, in pharmaceutical products containing licorice. The goal is to create and verify an HPLC-DAD technique that will enable the simultaneous study of GB and GA without affecting other elements. The suggested approach will make it simple and quick to analyze pharmaceutical goods that are purported to contain GB, GA and licorice in standard laboratories.

2. Materials and methods

2.1 Chemicals and reagents

Acetonitrile, methanol, ethanol, chloroform, dimethylsulfoixde, triethylamine, glycyrrhizic acid and glabridin (≥95.0% HPLC) were acquired from Sigma-Aldrich. Merck provided the monobasic potassium phosphate (KH_2PO_4). About 0.45-μm polyethersulfone filters (Dainippon Seiki, Kyoto, Japan) were used for sample filtration. A human system (Japan) provided water purification.

2.2 Preparation of mobile phase and solutions

Monobasic potassium phosphate buffer (1.36 mg/mL) was included in mobile phase A. There was 100% acetonitrile in mobile phase B. As a diluent, 100% methanol was utilized. Making mixed standard solutions and stock solutions of glabridin and glycyrrhizic acid in order to create GA solutions of 0.1 mg/mL, 2.6 mg of the glycyrrhizic acid reference standard was dissolved in 25 mL of diluent, and 2.5 mg of the glabridin reference standard was dissolved in 25 mL of diluent. In order to create a mixed standard solution, 20 mL of methanol was used to dilute 2 mL of each standard stock solution—glabridin and glycyrrhizic acid—to get a concentration of 0.01 mg/mL for GA and GB. Working solutions were prepared by dilutions of various kinds. The stock standard and working solutions were kept +4°C and were remained stable throughout the research. For HPLC analysis, working standard solutions were aliquoted in quantities of 30 μL. The chromatograms were analyzed using the peak regions versus concentration of the medicinal compounds.

2.3 Selection of wavelength

Spectrophotometric measurements were carried out using a Shimadzu UV-160 A spectrophotometer with 1-cm glass cells. The scanned UV spectra of the GA and GB solution, which ranged from 200 to 400 nm, were used to determine the detection of wavelength. The selected wavelength for GB was 280 nm, whereas the wavelength for GA was 254 nm, based on the UV spectrum.

2.4 Instrumentation and chromatographic conditions

The Shimadzu (Japan) liquid chromatography system, which included an LC-20 AT pump, SIL AT-HT autosampler component, SPD-20A HT DAD detector set at 254 nm and 280 nm, and CTO 10 AC column oven, was utilized for the HPLC analyses. To achieve the most effective chromatographic separation, a range of mobile phase, column types and stationary phase size combinations were tested with varying flow rates and column temperatures. On a GL Sciences (Japan) C18 (ODS) column with the dimensions of 4.6 mm I.D., 150 mm length and 5 μm particle size, chromatographic separation was achieved isocratically at 30°C. KH_2PO_4 makes up mobile phase A, while acetonitrile (100%), with a flow rate 1.3 mL/min, makes up mobile phase B.

2.5 Preparation of the calibration curves

A calibration curve was developed by examining working standard solutions of GA and GB at various concentrations. The linear concentration ranges of the method for the two medicines were calculated using linear least-squares regression (each concentration was examined as five repetitions). The calibration curve's equations have been obtained as $y = ax + b$, where x represents the drug substance concentrations in μg/mL and y represents the peak areas.

2.6 Application of the extraction process from the analyzed pharmaceutical products

Different extraction methods, such as liquid-liquid extraction (LLE) and solid phase extraction (SPE), are established for the purpose of extracting GA and GB out of pharmaceutical products. The liquid-liquid extraction (LLE) approach was initially investigated using a range of extraction solvents, combinations and extraction solvent volumes to create an efficient method of extracting GA and GB from pharmaceutical products employing the solid phase extraction (SPE) technique, and experiments were conducted using various elution procedures employing reversed phase and normal phase sorbents. Some of the solvent systems tested for liquid-liquid extraction are methanol-water, ethanol-water, methanol-acetonitrile, ethanol-acetonitrile, chloroform-water, chloroform-acetonitrile, dimethylsulfoxide-water and dimethylsulfoxide-acetonitrile. Following these trials, which demonstrated higher recovery values and increased efficiency from LLE, the following procedure was chosen to be used. These conditions given below are the final liquid-liquid extraction conditions and are used to extract GA and GB from pharmaceutical formulations at the highest yield.

About 0.5 g of pharmaceutical samples was weighed, and 1.5 mL of an ethanol: acetonitrile (1:1) mixture was added to the Eppendorf tube. After 30 seconds of vortexing, the mixture went through a centrifuge for 5 minutes at 4000 rpm. Using 0.45-μm polyethersulfone filters, 800 μL of the supernatant was filtered, and the liquid was then transferred to a 1.5 mL HPLC vial. For every duplicate, 30 μL of the liquid from the vial was administered. The nominal contents of pharmaceutical products were calculated using the regression equations derived from the calibration graphs.

3. Results and discussion

3.1 Method development

The best chromatographic conditions were found by conducting exploratory tests using the HPLC technique. Different temperatures were used to examine different kinds of columns. The best resolution, with more symmetrical and sharper peaks, was obtained using a C18 column that had the following specifications: 250 mm in length, 4.6 mm in diameter and 5 μm particle size, all maintained at 30°C. At varied flow rates, many mobile phases—including both aqueous and acidic solutions—were examined. It was found that the acidic mobile phase produced positive outcomes. As a result, the acidic component was chosen to be a monobasic potassium phosphate solution containing 1 mL/L triethylamine. The organic modifier of choice was acetonitrile. Acetonitrile-monobasic potassium phosphate (pH: 4.4) with a ratio of 90:10 (v/v) at a flow rate of 1.3 mL/min was found to be the ideal mobile phase composition, producing a high-resolution value. We chose two wavelengths, 254 and 280 nm, for quantification. The cosmetic raw material retention durations under these conditions were determined to be 6.10 ± 0.01 for GA and 8.62 ± 0.01 for GB. The ideal conditions for analysis were determined by measuring peak areas and resolution values. The chromatograms in **Figure 4(a–c)** are representative examples. In the previous study, the retention time for GA and GB was 32 and 51 minutes, respectively. This highlights our method as a quality parameter [23].

3.2 Method validation

The International Conference on Harmonization's (ICH) requirements were followed in the validation of the approach [26, 27].

3.2.1 Linearity range and calibration curve

For GA and GB, the linearity ranges for the analysis were set at 0.01–30 μg/mL and 0.05–20 μg/mL, respectively. The following regression equations were found for the calibration curves:

For GA: $y = 104{,}814x - 18{,}554$ (with an r^2 value of 0.9992)

For GB: $y = 51{,}060x - 3200$ (with an r^2 value of 0.9995)

Using the formulae LOD or LOQ = k * SDa/b—where k = 3 for LOD and 10 for LOQ—the values of the limits of detection (LOD) and quantification (LOQ) were calculated. Sda stands for the standard deviation of the intercept, while b is the slope. The LOD values in this case were found to be 0.016 μg/mL for GB and 0.003 μg/mL for GA. Similarly, it was found that the LOQ values for GA and GB were 0.01 and

DOI: http://dx.doi.org/10.5772/intechopen.1006964

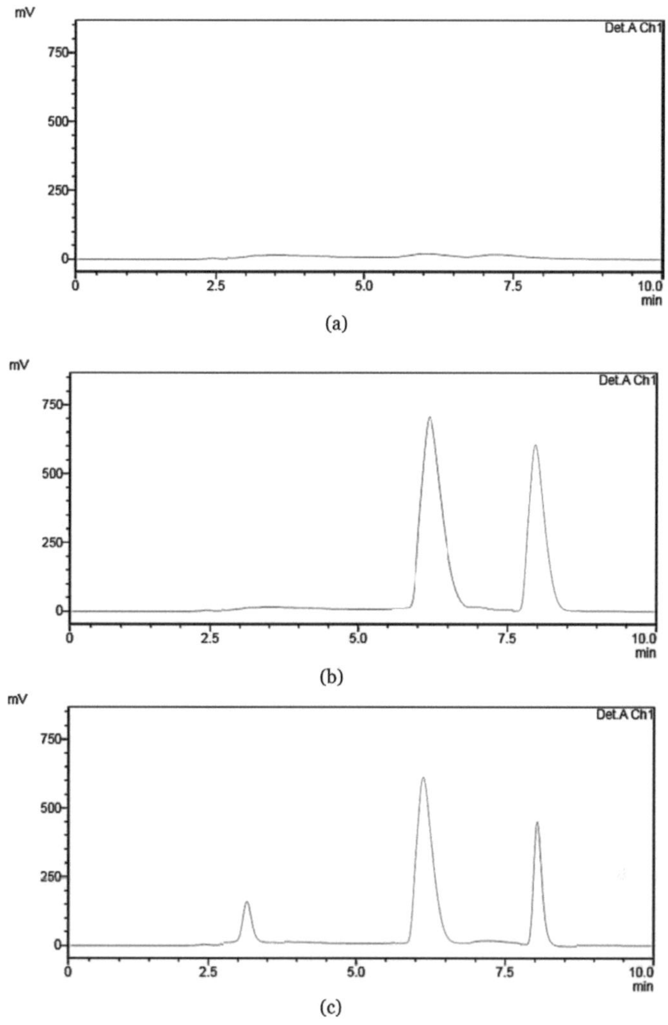

Figure 4.
a: Blank (aqueous medium), b: standard solution (10 μg/mL standard GA and GB solutions), c: real sample (nasal spray).

0.05 μg/mL, respectively. **Table 1** contains further analytical parameters. Although the previous study used a mass spectrometer detector, the LOD value was 0.95 μg/mL, while in our study it was found to be 0.003 μg/mL for GA. This shows that our method is more sensitive [25].

3.2.2 Precision

Three concentration levels (low, medium and high) were used to assess the method's precision in terms of both hourly and daily precisions. Both GA and GB underwent these analyses on the same day as well as seven distinct days, with five runs of each test. For hourly precision, the relative standard deviation (RSD) values varied from 0.25 to 1.57%, and for daily precision, they ranged from 1.30 to 2.85%. These outcomes show how highly repeatable the procedure is.

Parameter	GA	GB
Linearity range* (µg/mL)	0.01–30	0.05–20
Regression equation	y = 104,814x – 18,554	y = 51,060x – 3200
Slope ± SD	104,814 ± 80	51,060 ± 55
Intercept±SD	18,554 ± 35	3200 ± 22
Correlation coefficient, r^2	0.9992	0.9995
LOD (µg/mL)	0.003	0.016
LOQ (µg/mL)	0.01	0.05

**n = 5 correspond to replicate analysis for each level.*

Table 1.
Results of analytical parameters for the proposed method.

3.2.3 Recovery

To ascertain the precision of the technique for quantifying GA and GB, recovery experiments were conducted. The following equation was employed for the recovery trials, and the method's recovery was verified at three different concentrations by spiking the standards to the pharmaceutical formulations listed in **Table 2**.

$$\text{Recovery\%} = \left(\left(\text{Ct} - \text{Cu}\right)\right) / \text{Ca}) \times 100 \qquad (1)$$

In the formula, the percentage of the analyte that has been purified is denoted by Ca, the percentage of the analyte that is contained in the formula by Cu, and the total concentration of the analyte that has been found by Ct. The recovery study's results and the industrial pharmaceutical formulation specimen kinds' analysis are presented in **Table 2**. A quantitative range of 97.00–99.01% was seen in the average percent recoveries obtained from the approach. The recoveries in these data are extremely near to 100%, indicating a high degree of accuracy in the approach and a good level of dependability for figuring out the analyte concentration in pharmaceutical

	Existing concentration (µg/mL)	Added concentration (µg/mL)	Found concentration* (mg/mL) (mean ± S.D)	Recovery (%)	RDS of intraday variation	RDS of interday variation
		0.01	1.00 ± 0.02	99.01	0.25	1.30
GA	1	1	1.94 ± 0.04	97.00	1.57	2.85
		30	30.44 ± 0.03	98.19	1.44	2.13
		0.05	1.03 ± 0.02	98.09	1.35	2.06
GB	1	1	1.96 ± 0.03	98.00	1.51	2.57
		20	20.73 ± 0.01	98.71	1.12	2.40

**n = 5 correspond to replicate analysis for each level.*

Table 2.
Results of recovery studies by standard addition method.

formulations. In the previous study, the RSD values were considerably higher than our study, indicating that the reproducibility of the method was low. In addition, all our recovery values were found to be above 95% [24].

3.2.4 Stability

Under a variety of storage settings, such as room temperature in the dark, autosampler conditions for 48 hours and refrigeration at 4°C for a month, the stability of the working standard solutions of the pharmaceutical raw materials was evaluated. The samples continued to be stable in these circumstances, according to the stability study results. In particular, it was discovered that the samples were stable after 48 hours of room temperature storage and the same amount of time under autosampler settings. Additionally, the samples' stability was unaffected by a month of chilling at 4°C. Both GA and GB were shown to be stable under all evaluated storage settings, indicating that sample degradation during storage need not be a worry for the method's ability to accurately assess these pharmaceutical raw materials.

3.2.5 Robustness

By altering important variables such as the flow rate, temperature of the column oven, and the ratios of acetonitrile and the acidic solution in the mobile phase, the robustness of the procedure was assessed. Two distinct ratios—85:15 and 95:5—were substituted for the original 90:10 (acetonitrile-monobasic potassium phosphate solution) composition of the mobile phase. The flow rate was changed from 1.3 to 1.2 and 1.4 mL/min, and the column temperatures were changed from the starting 30–35 and 25°C. Notably, these adjustments to the method's parameters had no discernible effect on the peak regions. **Table 3** shows low relative standard deviation (RSD) values, which shows how reliable the approach is. This implies that the approach may yield consistent and dependable findings even in the presence of small fluctuations in these parameters, which is a crucial feature for its practical usefulness.

3.3 Analysis of real samples

The concentrations of GA and GB in pharmaceutical goods containing *Glycyrrhiza glabra* L. were obtained following the technique and validation experiments and are presented in **Table 4**. The percentage RSD values were quite low in comparison with

Condition	Value	Recovery (%)		RSD (%)	
		GA	GB	GA	GB
Flow rate mL/min	1.2	98.53	99.88	0.24	0.17
	1.4	99.32	99.19	0.29	0.27
Mobile phase composition (ACN:aqueous phase)	85:15	98.96	98.89	1.15	1.21
	95:5	97.15	97.80	1.68	1.61
Column temperature	25	99.16	99.86	0.22	0.16
	35	99.64	99.75	0.31	0.28

For each concentration n = 5.

Table 3.
Results from robustness experiments.

Cosmetic product	Recovery	%RSD	GA (μg/mL)	GB (μg/mL)
Body spray	% 97.8	0.84	24.18	17.98
Cream	% 96.6	1.13	22.97	16.41
Lotion	% 96.4	1.17	22.12	16.04
Ointment	% 95.1	1.65	20.83	14.92
Lozenge	% 95.6	1.54	21.68	15.78
Nasal spray	% 98.1	0.67	26.33	18.44
Nasal ointment	% 97.4	0.91	23.44	17.21
Ear drops	% 92.6	2.24	17.46	13.07
Soap	% 94.4	1.92	19.43	14.01
Shampoo	%93.6	2.10	18.74	13.72

For each concentration n = 5.

Table 4.
Pharmaceutical raw material concentrations and method reproducibility.

our previous work [28]. %RSD values ranged from 0.67 to 2.24. While the amount of GA varied between 26.33 and 18.74 μg/mL, the amount of GB varied between 18.44 and 13.72 μg/mL.

4. Conclusion

We devised a technique in this work to identify commonly used pharmaceutical formulations of GA and GB. GA and GB are utilized in a wide variety of pharmaceutical preparations because they have several medicinal and cosmetic benefits (as indicated by the text's references). The majority of GA and GB are present in *Glycyrrhiza glabra* L. sp. pharmaceutical drugs containing GA and GB are often introduced as preparations made of extract from *Glycyrrhiza glabra* L.

GA and GB are discovered in combination because of their comparable effects. In this regard, our goal was to create a novel HPLC technique that would allow us to determine the concentrations of GA and GB in 10 distinct pharmaceutical types, including nasal spray, shampoo, cream and so on. Our approach also saves costs, is incredibly easy to use and is quite quick. The validation experiments demonstrated the reasonable sensitivity, selectivity, precision and accuracy of this approach. The literature has a few techniques for determining GA and GB, but none of these methods allow analysis of products with such different types of pharmaceutical formulations. Current methods focus on plant extracts, dietary supplements and a single pharmaceutical formulation. In addition to the expensive detector requirements and costs of existing methods, the long retention times constitute a shortcoming. Simultaneous analysis of the maximum absorption of GA and GB at different wavelengths was carried out using the most economical diode array detector (DAD). The method we developed is superior to all methods available in the literature in terms of all validation parameters. The most important of these are retention times, detection limits, sensitivity, selectivity, RSD values and robustness. Retention times of 6.10 minutes for GA and 8.62 minutes for GB were not achieved even by ultra-high performance liquid chromatography. The LOD and LOQ values obtained using the DAD detector could not be obtained even using the MS detector. Since pharmaceutical formulations are different and unique for each product, there is no cost-effective separation method that removes active ingredients from their environment with such a simple and single extraction method. It also stands out as an environmentally friendly method due to the low amount of solvent used. Moreover, there is no simple extraction method to extract the active ingredients from pharmaceutical products with such different formulations. The approach is anticipated to be useful for pharmaceutical product standardization, quality control and routine analysis, including GA and GB.

Conflict of interest

The author declares no conflict of interest.

Author details

Burhan Ceylan
Department of Pharmacognosy, Faculty of Pharmacy, Harran University, Haliliye, Sanlıurfa, Turkey

*Address all correspondence to: b.ceylan022@gmail.com; burhanceylan@harran.edu.tr

References

[1] Casulli F, Ippolito A. Observations on liquorice rust (*Uromuces glycyrrhizae* (Rab.) Magn.) in Southern Italy. Informatore Fitopatologico. 1995;**45**:27-30

[2] Tian M, Yan H, Row KH. Extraction of glycyrrhizic acid and glabridin from licorice. International Journal of Molecular Sciences. 2008;**9**:571-577. DOI: 10.3390%2Fijms9040571

[3] Asl MN, Sosseinzadeh H. Review of pharmacological effects of *Glycyrrhiza sp*. and its bioactive compounds. Phytotherapy Research. 2008;**22**:709-724. DOI: 10.1002/ptr.2362

[4] Zhang Q, Ye M. Chemical analysis of the Chinese herbal medicine Gan-Cao (licorice). Journal of Chromatography A. 2009;**13**:1954-1969. DOI: 10.1016/j.chroma.2008.07.072

[5] Dudics S, Meka LRR, Venkatesha SH, Berman BM, Che CT, Moudgil KD. Natural products for the treatment of autoimmune arthritis: Their mechanisms of action, targeted delivery, and interplay with the host microbiome. International Journal of Molecular Sciences. 2018;**19**:2508. DOI: 10.3390%2Fijms19092508

[6] Matsumoto T, Tanaka M, Yamada H, Cyong JC. Effect of licorice roots on carrageenan-induced decrease in immune complexes clearance in mice. Journal of Ethnopharmacology. 1996;**26**:1-4. DOI: 10.1016/0378-8741(96)01415-8

[7] Chen M, Theander TG, Christensen SB, Hviid L, Zhai L, Kharazmi A. Licochalcone A, a new antimalarial agent, inhibits in vitro growth of the human malaria parasite plasmodium falciparum and protects mice from P. Yoelii infection. Antimicrobial Agents and Chemotherapy. 1994;**38**:1470-1475. DOI: 10.1128/aac.38.7.1470

[8] Friss-Moller A, Chen M, Fuursted K, Christensen SB, Kharazmi A. In vitro antimycobacterial and antilegionella activity of licochalcone a from Chinese licorice roots. Planta Medica. 2002;**68**:416-419. DOI: 10.1055/s-2002-32087

[9] Fukai T, Marumo A, Kaitou K, Kanda T, Terada S, Nomura T. Antimicrobial activity of licorice flavonoids against methicillin-resistant *Staphylococcus aureus*. Fitoterapia. 2002;**73**:536-539. DOI: 10.1016/s0367-326x(02)00168-5

[10] Klonoff DC. Chronic fatigue syndrome and neurally mediated hypotension. Journal of the American Medical Association. 1996;**275**:359-360

[11] Takahashi K, Yoshino K, Shirai T, Nishigaki A, Araki Y, Kitao M. Effect of a traditional herbal medicine (shakuyaku-kanzo-to) on testosterone secretion in patients with polycystic ovary syndrome detected by ultrasound. Nihon Sanka Fujinka Gakkai Zasshi. 1988;**40**:789-792

[12] Takahashi K, Kitao M. Effect of TJ-68 (shakuyaku-kanzo-to) on polycystic ovarian disease. International Journal of Fertility and Menopausal Studies. 1994;**39**:69-76

[13] Fu Y, Hsieh T, Guo J, Kunicki J, Lee MYWT, Darzynkiewicz Z, et al. Licochalcone-A, a novel flavonoid isolated from licorice root (*Glycyrrhiza glabra*), cause G2 and late-G1 arrests in androgen-independent PC-3 prostate cancer cells. Biochemical and Biophysical

Research Communications. 2004;**10**:263-270. DOI: 10.1016/j.bbrc.2004.07.094

[14] Saeedi M, Morteza-Semnani K, Ghoreishi MR. The treatment of atopic dermatitis with licorice gel. Journal of Dermatological Treatment. 2003;**14**:153-157. DOI: 10.1080/09546630310014369

[15] Eichenfield LF, Fowler JF, Rigel DR, Taylor SC. Natural advances in eczema care. Cutis. 2007;**80**:2-16

[16] Mukhopadhyay M, Panja P. A novel process for extraction of natural sweetener from licorice (*Glycyrrhiza glabra*) roots. Separation and Purification Technology. 2008;**63**:539-545. DOI: 10.1016/j.seppur.2008.06.013

[17] Kiso A, Kawashima Y, Ohto N, Zhou YY, Yashiki K, Murakami T, et al. Abstracts: The effects of licorice leaf extract on ceramide and hyaluronan synthesis. International Journal of Cosmetic Science. 2010;**43**:391-392. DOI: 10.1111/j.1468-2494.2010.00583_4.x

[18] Van Rossum TGJ, Vulto RA, Man D, Schalm B. Glycyrrhizin as a potential treatment for chronic hepatitis C. Alimentary Pharmacology and Therapeutics. 2001;**12**:199-205. DOI: 10.1046/j.1365-2036.1998.00309.x

[19] Kwon HS, Oh SM, Kim JK. Glabridin, a functional compound of liquorice, attenuates colonic inflammation in mice with dextran sulphate sodium-induced colitis. Clinical and Experimental Immunology. 2008;**151**:165-173. DOI: 10.1111%2Fj.1365-2249.2007.03539.x

[20] Yokota T, Nishio H, Kubota Y, Mizoguchi M. The inhibitory effect of glabridin from licorice extracts on melanogenesis and inflammation. Pigment Cell Research. 1998;**11**:355-361. DOI: 10.1111/j.1600-0749.1998.tb00494.x

[21] Capella GL, Finzi AF. Complementary therapy for psoriasis. Dermatologic Therapy. 2003;**16**:164-174. DOI: 10.1046/j.1529-8019.2003.01625.x

[22] Therapeuticresearchfaculty TRF. Natural Medicines Comprehensive Database. 2015. Available from: http://naturaldatabase.therapeuticresearch.com

[23] Kulkarni S, Palghadmal S, Sawant A, Bhope S, Padmanabhan S. RP-HPLC method development and validation for simultaneous estimation of glabridin, glycyrrhiziniz acid and total flavonoids in *Glycyrrhiza* glabra extract: Stability studies of constituents. Global Journal of Pharmacy and Pharmaceutical Sciences. 2021;**8**:555738

[24] Kumbhalkar B, Walunj T, Chavan S, Gupta V, Deshmukh V, Sardeshmukh S. Simultaneous identification and estimation of glycyrrhizin, glabridin and 18β-glycyrrhetinic acid in de-glycyrrhized Ayurvedic lipid-based formulation of *Glycyrrhiza glabra* using dual wavelength reverse phase-high-performance liquid chromatography. Seperation Science Plus. 2020;**3**:460-471. DOI: 10.1002/sscp.202000015

[25] Li G, Nikolic D, Breemen RB. Identification and chemical standardization of licorice raw materials and dietary supplements using UHPLC-MS/MS. Journal of Agricultural and Food Chemistry. 2016;**64**:8062-8070. DOI: 10.1021%2Facs.jafc.6b02954

[26] Guidance for Industry, Bioanalytical Method Validations. Rockville MD: US Department of Health and Human Services, Food and Drug Administration, CDER; 2001. DOI: 10.1016/j.ica.2020.119681

[27] Text on Validation of Analytical Procedures Q2A, International Conference on Harmonisation of Technical Requirements for Registration of Pharmaceuticals for Human Use. London; 1994

[28] Serim E, Ceylan B, Tekkeli-Kepekci SE. Determination of apigenin in cosmetic containing chamomile by high-performance liquid chromatography with ultraviolet detection (HPLC-UV). Analytical Letters. 2023;**56**:2113-2122. DOI: 10.1080/00032719.2022.2155180